J S Burgler

Cardiovascular Physiology

Lois Jane Heller, Ph.D.

David E. Mohrman, Ph.D.

Department of Physiology
School of Medicine
University of Minnesota, Duluth

McGraw-Hill Book Company

New York St. Louis San Francisco Auckland Bogotá
Guatemala Hamburg Johannesburg Lisbon London Madrid
Mexico Montreal New Delhi Panama Paris San Juan
São Paulo Singapore Sydney Tokyo Toronto

NOTICE

Medicine is an ever-changing science. As new research and clinical experience broaden our knowledge, changes in treatment and drug therapy are required. The editors and the publisher of this work have made every effort to ensure that the drug dosage schedules herein are accurate and in accord with the standards accepted at the time of publication. Readers are advised, however, to check the product information sheet included in the package of each drug they plan to administer to be certain that changes have not been made in the recommended dose or in the contraindications for administration. This recommendation is of particular importance in regard to new or infrequently used drugs.

This book was set in Press Roman by Hemisphere Publishing Corporation. The editor was Richard W. Mixter and the production supervisor was Carol Swain.
The cover was designed by Jane Moorman. The drawings were done by ANCO/Boston.
Fairfield Graphics was printer and binder.

CARDIOVASCULAR PHYSIOLOGY

1 2 3 4 5 6 7 8 9 0 FGFG 8 9 8 7 6 5 4 3 2 1 0

Library of Congress Cataloging in Publication Data

Heller, Lois Jane.
Cardiovascular physiology.

Bibliography: p.
Includes index.
1. Cardiovascular system. I. Mohrman, David E., joint author. II. Title.
[DNLM: 1. Cardiovascular system–Physiology. WG102 H477c]
QP102.H44 612'.1 80-21277
ISBN 0-07-027973-X

Contents

Preface

This text is intended to provide first-year medical students with the core of information and concepts necessary to develop a firm understanding of how the intact cardiovascular system operates. Specifically stated learning objectives and study questions for each chapter allow the student to test his or her mastery of the material presented. This format lends itself to independent study, which may (but need not) be supplemented by additional lecture material. References are supplied for each chapter to provide interested students with access to the pertinent research literature.

We feel strongly that cardiovascular instruction in the first-year medical curriculum should give the student not simply a collection of facts but also an understanding of how the intact cardiovascular system operates. Cardiovascular physiology is often a student's first exposure to the operation of a complete organ system, and the student therefore often finds it confusing to deal with the continual interactions that occur among the various system components. Consequently, we have tried to direct our presentation throughout toward the overall operation of the cardiovascular system rather than attempting to present all available facts. We would welcome your comments and criticisms of this approach.

We wish to express our thanks to our mentors, who by example have provided us with the incentives and enthusiasm to undertake this project; to the University of Minnesota at Duluth students, who have evaluated our efforts and supplied us with constructive criticism; to Carol Peterman, who took care of the preparation of the manuscript; and especially to our spouses, Bob and Peggy, who provided essential moral support.

Lois Jane Heller
David E. Mohrman

Chapter 1

Homeostasis and Cardiovascular Transport

OBJECTIVES

The student understands the basic principles of cardiovascular transport and their role in maintaining homeostasis:

1. Defines homeostasis.
2. Identifies the major body fluid compartments and states the approximate volume of each.
3. Diagrams the blood flow pathways between the heart and other major body organs.
4. Lists the two factors, provided by the cardiovascular system, that are essential for regulating the composition of interstitial fluid by transcapillary exchange.
5. States the relationship among blood flow, blood pressure, and vascular resistance.
6. Predicts the percentage change in flow through a tube caused by a doubling of tube length, tube radius, fluid viscosity, or pressure difference.

7 Defines bulk transport and diffusion and lists the factors that determine the rate of each.
8 Given data, uses the Fick principle to calculate the rate of removal of a solute from blood as it passes through an organ.
9 Describes how capillary wall permeability to a solute is related to the size and lipid solubility of the solute.
10 Lists the factors that influence transcapillary fluid movement and, given data, predicts the direction of transcapillary fluid movement.
11 Describes the lymphatic vessel system and its role in preventing fluid accumulation in the interstitial space.

DEFINITION OF HOMEOSTASIS

A nineteenth century French physiologist, Claude Bernard, first recognized that all higher organisms actively and constantly strive to prevent the external environment from upsetting the conditions necessary for life within the organism. Thus the temperature, oxygen concentration, pH, ionic composition, osmolarity, and many other components of our *internal environment* are closely controlled. This process of maintaining the constancy of our internal environment has come to be known as *homeostasis.* To accomplish this task, an elaborate material transport network, the cardiovascular system, has evolved.

Various compartments of watery fluids, known collectively as the *total body water*, account for about 60 percent of body weight. This water is distributed among the *intracellular, interstitial,* and *plasma* spaces as indicated in Fig. 1-1. About two-thirds of our body water is contained within cells and communicates with the interstitial fluid across the plasma membranes of cells. Of the fluid that is outside cells, only a small amount, the *plasma volume*, circulates within the cardiovascular system. The circulating plasma fluid communicates with the interstitial fluid across the walls of small capillary vessels.

The interstitial fluid is the immediate environment of individual cells. These cells must draw their nutrients from and release their products into the interstitial fluid. The interstitial fluid cannot, however, be considered as a large reservoir for nutrients or a large sink for metabolic products since its volume is less than half that of the cells that it serves. The well-being of individual cells therefore depends heavily on the homeostatic mechanisms that regulate the composition of the interstitial fluid. This task is accomplished by continuously exposing the interstitial fluid to "fresh" circulating plasma fluid.

As the blood passes through capillaries, solutes exchange between it and the interstitial fluid by the process of diffusion. The net result of transcapillary diffusion is always that the interstitial fluid tends to take on the composition of the incoming blood. If, for example, the potassium-ion concentration in the interstitium of a particular skeletal muscle were higher than that in the blood entering the muscle, potassium would diffuse into the blood as it passed through

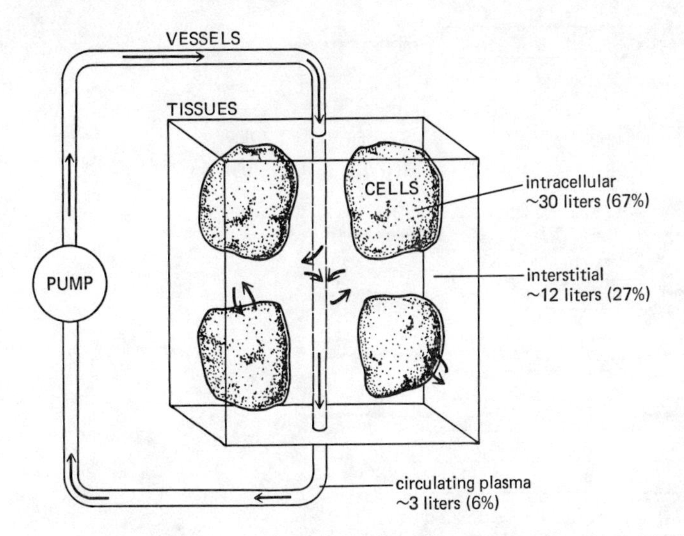

Figure 1-1 Major body fluid compartments with average volumes indicated for a 70-kg man. Total body water is about 60 percent of body weight. Numbers in parentheses indicate approximate percentage of total body water in each compartment.

the muscle's capillaries. Since this removes potassium from the interstitial fluid, the interstitial potassium-ion concentration would decrease. It would stop decreasing when net movement of potassium into capillaries no longer occurred; i.e., when the interstitial concentration reached that of the incoming plasma.

Two conditions are essential for this circulatory mechanism to effectively control the composition of interstitial fluid: (1) there must be adequate blood flow through the tissue capillaries, and (2) the chemical composition of the incoming (or arterial) blood must be controlled to be that which is desired in the interstitial fluid. These conditions are met by the design and operation of the cardiovascular system.

COMPONENTS OF THE SYSTEM

The overall functional arrangement of the cardiovascular system is illustrated in Fig. 1-2. Since a functional rather than an anatomic viewpoint is expressed in this figure, the heart appears in three places: as the right heart pump, as the left heart pump, and as the heart muscle tissue. It is common practice to view the cardiovascular system as (1) the *pulmonary circulation*, composed of the right heart pump and the lungs, and (2) the *systemic circulation*, in which the left heart pump supplies blood to the systemic organs (all structures except the gas exchange portion of the lungs).

When blood leaves the left heart, it passes through only one of the systemic organs since their circulations are generally arranged in parallel. There are two

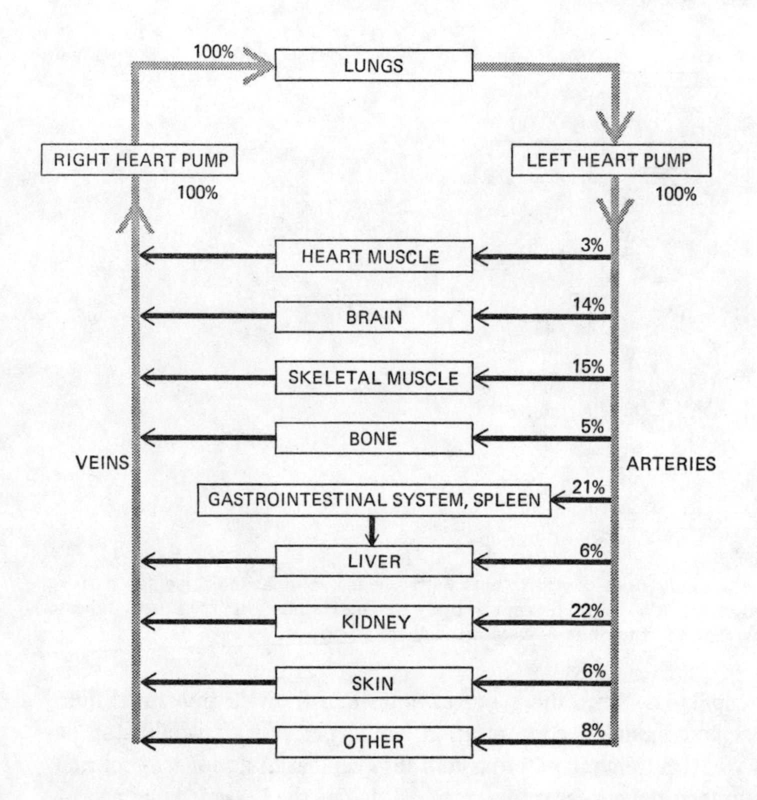

Figure 1-2 Cardiovascular circuitry indicating percentage distribution of cardiac output to various organ systems.

important results of this parallel arrangement. First, all systemic organs receive blood of identical composition—that which has just left the lungs and is known as *arterial blood*. Second, the flow through any one of the systemic organs can be controlled independently of the flow through the other organs. Thus, for example, skeletal muscle flow can increase severalfold, as it does during exercise, without necessarily influencing flow through the brain.

Many of the organs in our bodies help perform the task of continually reconditioning the blood circulating in the cardiovascular system. Key roles are played by organs such as the lungs, which communicate with the external environment. As is evident from the arrangement shown in Fig. 1-2, any blood that has just passed through a systemic organ returns to the right heart and is pumped through the lungs, where oxygen and carbon dioxide are exchanged. Thus the blood's gas composition is always reconditioned immediately after leaving a systemic organ.

Like the lungs, many of the systemic organs also serve to recondition the composition of blood, although the flow circuitry precludes their doing so each time the blood completes one circuit. The kidneys, for example, continually adjust the electrolyte composition of the blood passing through them. Because the blood conditioned by the kidneys mixes freely with all the circulating blood and because electrolytes and water freely pass through most capillary walls, the kidneys control the electrolyte balance of the entire internal environment. To achieve this, it is necessary that a given unit of blood pass often through the kidneys. In fact, the kidneys (under resting conditions) normally receive about one-fourth of the blood that leaves the left heart each minute, i.e., one-fourth of the *cardiac output*. This greatly exceeds the amount of flow that is necessary to supply the nutrient needs of the renal tissue. This situation is common to organs that have a blood-conditioning function.

Blood-conditioning organs can also withstand, temporarily at least, severe reductions of blood flow. Skin, for example, can tolerate restricted blood flow when it is necessary for the body to conserve heat. Most of the large abdominal organs also fit in this category. The reason is simply that because of their blood-conditioning functions, their normal blood flow is far in excess of that necessary to maintain their basal metabolic needs.

The brain, heart muscle, and skeletal muscles typify organs in which blood flows solely to supply the metabolic needs of the tissue. They do not recondition blood for the benefit of any other organ. Flow to brain and heart muscle is normally only slightly greater than that required for their metabolism, and they do not tolerate blood flow interruptions well. Unconsciousness can occur within a few seconds after stoppage of cerebral flow, and permanent brain damage can occur in as little as 4 min without flow. Similarly, the myocardium normally consumes about 75 percent of the oxygen supplied to it, and the heart's pumping ability begins to deteriorate within beats of a coronary flow interruption.

BASIC PHYSICS OF CIRCULATION

Understanding the relationship between the physical factors that govern fluid flow in a tube is probably the most important key to comprehending how the cardiovascular system operates.

The tube depicted in Fig. 1-3 represents a segment of any vessel in the body.

Figure 1-3 Factors influencing fluid flow through a tube.

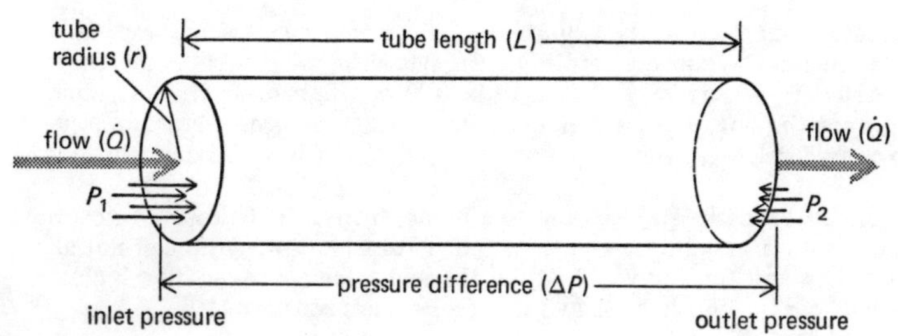

It has a certain length (L) and a certain internal radius (r) through which blood flows. Fluid flows through the tube only when the pressures in the fluid at either end (P_1 and P_2) are unequal, i.e., when there is a pressure difference (ΔP) between the ends. Pressure differences supply the driving force for flow. Flow is driven by pressure differences in the same sense that diffusion is driven by concentration differences. Because friction develops between the moving fluid and the stationary walls of a tube, vessels tend to resist fluid movement through them. This *vascular resistance* is a measure of how difficult it is to cause fluid to flow through the tube, i.e., how much of a pressure difference it takes to cause a certain flow. The all-important relation among flow, pressure difference, and resistance is described by the *basic flow equation* as follows:

$$\text{Flow} = \frac{\text{pressure difference}}{\text{resistance}}$$

$$\dot{Q} = \frac{\Delta P}{R}$$

where $\dot{Q}$ = flow rate (volume/time)
ΔP = pressure difference (mmHg[1])
R = resistance to flow (mmHg × time/volume)

It should be noted that this relationship may be applied not only to a single tube but to collections of tubes, e.g., to the vascular bed of an organ or to the entire systemic system. Furthermore, it should be evident from the basic flow equation that there are only two ways in which blood flow through any organ can be changed: (1) by changing the pressure difference across its vascular bed, or (2) by changing its vascular resistance. Most often, it is changes in an organ's vascular resistance that cause the flow through the organ to change.

Resistance to flow of fluid through a tube depends on several factors, including the radius and length of the tube and the viscosity of the fluid flowing through it. The following equation represents an expansion of the basic flow equation to include these factors and is known as Poiseuille's equation.[2]

$$\dot{Q} = \Delta P \frac{\pi r^4}{8L} \frac{1}{\eta}$$

[1] Although pressure is most correctly expressed in units of force per unit area, it is customary to express pressures within the cardiovascular system in millimeters of mercury. For example, mean arterial pressure may be said to be 100 mmHg because it is the same as the pressure existing at the bottom of a mercury column 100 mm high. All cardiovascular pressures are expressed relative to atmospheric pressure, which is approximately 760 mmHg.

[2] Poiseuille's equation properly applies only to a homogeneous fluid flowing through rigid nontapered tubes with a certain flow pattern called laminar flow. Although not all these conditions are rigidly met for any vessel within the body, the approximation is close enough to permit general conclusions to be drawn from Poiseuille's equation.

where r = inside radius of the tube
L = tube length
η = fluid viscosity
$\dot{Q}$ and ΔP as previously defined

The obvious conclusion to be drawn from the Poiseuille equation is that internal radius has a tremendous effect on a tube's resistance to flow. Because it is raised to the fourth power, decreasing the inside radius to one-half will decrease the flow caused by a given pressure difference 16-fold. It is logical that the vascular resistance (and therefore blood flow) of an organ is regulated primarily by changing the inside radius of vessels within the organ. Whereas vessel length and blood viscosity are factors that affect vascular resistance, they are not variables easily manipulated for the purpose of controlling blood flow.

CARDIOVASCULAR TRANSPORT

The cardiovascular system is a network for moving substances from one location to another. Its efficient design permits it to effectively control the chemical composition of the entire internal environment using a very limited volume of circulating fluid. The cardiovascular system operates on only two basic physical principles: *bulk transport*, by which substances get from the capillary bed of one organ to that of another organ by being swept along with the fluid in which they are contained, and *diffusion*, by which substances enter or leave the circulating fluid across the capillary vessel wall according to the direction of the concentration gradient.

Bulk Transport and the Fick Principle

The rate at which a substance (X) moves from one place to another within the vascular system depends solely on the concentration of the substance in blood and the blood flow rate as follows:

$$\text{Transport rate} = \text{flow rate} \times \text{concentration}$$

or

$$\dot{X} = \dot{Q}[X]$$

where $\dot{X}$ = rate of transport of X (mass/time)
$\dot{Q}$ = blood flow rate (volume/time)
[X] = concentration of X in blood (mass/volume)

It is clear from the equation above that only two means are available for altering the rate at which a substance is carried to an organ: (1) a change in the blood flow rate through the organ, or (2) a change in the arterial blood concentration of the substance. The equation above might be used, for example, to calculate how much oxygen is carried to a certain skeletal muscle each minute. Note, however, that this calculation would not indicate whether the muscle actually used the oxygen carried to it.

One can expand the bulk transport principle to determine a tissue's rate of utilization of a substance by simultaneously considering the transport rate of the substance to *and from* the tissue. The relationship that results is referred to as the Fick principle and may be formally stated as follows:

$$\dot{X}_{tc} = \dot{Q}([X]_a - [X]_v)$$

where $\dot{X}_{tc}$ = transcapillary efflux rate of X (mass/time)
$\dot{Q}$ = flow rate (volume/time)
$[X]_{a,v}$ = arterial and venous concentrations of X

The Fick principle essentially says that the amount of a substance that goes into an organ in a given period of time ($\dot{Q}[X]_a$) minus the amount that comes out ($\dot{Q}[X]_v$) must equal the tissue utilization rate of that substance. This principle finds many applications in physiological and clinical situations.

Transcapillary Solute Diffusion

Capillaries act as efficient exchange sites where most substances cross the capillary walls simply by *passively diffusing* from regions of high concentration to regions of low concentration. As in any diffusion problem, there are four factors that determine the diffusion rate of a substance between the blood and the interstitial fluid: (1) the concentration difference, (2) the surface area for exchange, (3) the diffusion distance, and (4) the specific permeability of the capillary wall to the diffusing substance.[3]

Capillary beds allow huge amounts of materials to enter and leave blood because they maximize the area across which exchange can occur while minimizing the distance over which the diffusing substances must travel. Capillaries are extremely fine vessels with a *lumen* (inside) diameter of about 7 μm, a wall thickness of approximately 1 μm, and a length of perhaps 0.5 mm. (For comparison, a human hair is roughly 100 μm in diameter.) Capillaries are distributed in incredible numbers in organs and communicate intimately with all regions of the interstitial space. It is estimated, for example, that a single cubic millimeter of heart muscle contains 1000 to 2000 individual capillaries. In the entire body, the capillary surface area across which diffusion exchange is continually occurring between blood and interstitial fluid is roughly 100 m^2! Duplicating this

[3] These factors are combined in the equation that describes the rate of diffusion ($\dot{X}_d$) of a substance X across a barrier: $\dot{X}_d = DA\,\Delta[X]/\Delta L$, where D, A, $\Delta[X]$, and ΔL represent the diffusion coefficient, surface area, concentration difference, and diffusion distance, respectively.

huge surface area and extremely short diffusion distance in an artificial lung or kidney is a difficult task indeed.

The capillary wall itself consists of only a single thickness of endothelial cells joined to form a tube. Careful experimental studies on how rapidly different substances cross capillary walls indicate that fundamentally distinct pathways exist for transcapillary exchange. Lipid-soluble substances, such as the gases oxygen and carbon dioxide, cross the capillary wall easily. This is predictable since the cell membranes that must be crossed present little barrier to small nonpolar particles.

Small polar particles such as sodium and potassium ions do not cross capillary walls as readily; we say that capillaries have a low permeability to these ions. Although low, the capillary permeability to small ions is several orders of magnitude higher than the permeability that would be expected if the ions were forced to move through the lipid plasma membranes. Indeed, equilibration of transcapillary ionic concentration differences occurs very rapidly in most tissues, indicating that the capillary wall is not a formidable barrier. It is therefore postulated that capillaries are somehow perforated at intervals with water-filled channels or *pores.* Calculations from diffusion data indicate that the collective cross-sectional area of the pores relative to the total capillary surface area varies greatly between capillaries in different organs. Brain capillaries appear to be very tight (have few pores), whereas capillaries in the kidney and fluid-producing glands are much more leaky. On the average, however, pores constitute only a very small fraction of total capillary surface area—perhaps 0.01 percent. An effective maximum diameter of about 40 Å has been assigned to individual pores since substances with molecular diameters larger than this essentially do not cross capillary walls. Thus albumin and other proteins in the plasma are confined to the plasma space.

TRANSCAPILLARY FLUID MOVEMENT

Water as well as solutes moves across capillary walls. Net shifts of fluid between capillary and interstitial compartments are of importance for a host of physiological functions, including maintenance of the circulating blood volume, intestinal fluid absorption, tissue edema formation, and saliva, sweat, and urine production. Intracapillary and interstitial fluid flows through the transcapillary channels in response to pressure differences across the capillaries in the same way that blood flows from high to low pressure along vessels. In fact, the hydrostatic pressure inside capillaries is generally higher than the interstitial or tissue pressure, so there is a strong tendency for fluid to flow out of capillaries through the pores—a process called *filtration.* Since intracapillary *hydrostatic pressure* is approximately 25 mmHg higher than the interstitial pressure in most tissues, our entire blood volume would soon be in the interstitial space if there were not some counteracting process tending to draw fluid into capillaries—a process known as *reabsorption*. The reabsorptive force is an *osmotic pressure* attributable to a higher concentration of protein in the plasma than in the interstitial fluid.

Recall that solvent always tends to move from regions of low to regions of high total solute concentration in establishing osmotic equilibrium. Also recall that osmotic forces are quantitatively expressed in terms of osmotic pressure. The osmotic pressure is defined for a given solution as the hydrostatic pressure necessary to prevent osmotic water movement when the test solution is exposed to pure water across a membrane permeable only to water. The osmotic pressure of a solution is proportional to the total number of solute particles in the solution.

Plasma, for example, has an osmotic pressure of about 5000 mmHg–nearly all of which is attributable to dissolved mineral salts such as NaCl and KCl. Since interstitial fluid also contains these salts at virtually identical concentrations, these particles do not affect transcapillary fluid movement. The osmotic gradient is generated by particles that are not capable of crossing the membrane barrier. These particles are primarily the protein molecules of the plasma and interstitial fluid.

Another term, *oncotic pressure*, is used to denote the portion of the plasma's osmotic pressure that is due to the presence of proteins, which do not freely cross capillaries. Oncotic pressure is proportional to protein concentration. Since plasma has a higher protein concentration than interstitial fluid, its oncotic pressure is higher. This difference creates the osmotic force that promotes fluid reabsorption into capillaries.

When the interstitial volume of a tissue is stable, as it usually is, there can be no net transcapillary fluid movement; i.e., the net 25-mmHg hydrostatic force for filtration must be exactly balanced by a net 25-mmHg oncotic force for fluid reabsorption. This delicate balance of the factors that determine transcapillary water movement is known as the Starling-Landis equilibrium[4] and may be expressed as follows:

$$\text{Net filtration rate} = K[(P_c - P_i) - (\pi_c - \pi_i)]$$

where P_c = the hydrostatic pressure of intracapillary fluid
π_c = the oncotic pressure of intracapillary fluid
P_i and π_i = the same quantities for interstitial fluid
K = a constant expressing how readily fluid can move across capillaries (essentially the reciprocal of the resistance to flow through the capillary wall)

[4] After the British physiologist, E. Starling, who proposed the idea and the U.S. physiologist, E. Landis, who experimentally validated it.

Equilibrium, or the absence of net transcapillary water movement, occurs when the bracketed term in this equation is zero. This equilibrium may be upset by alterations in any of the four pressure terms. For example, a substance called histamine is often released in damaged tissue. One of the actions of histamine is to increase capillary permeability to the extent that protein leaks into the interstitium. Net filtration and tissue swelling (edema) accompany histamine release, in part because the oncotic pressure difference $(\pi_c - \pi_i)$ is reduced below normal.

Transcapillary fluid filtration is not necessarily detrimental. Indeed, fluid-producing organs such as salivary glands and kidneys utilize high intracapillary pressure to produce continual net filtration. Moreover, in certain abnormal situations, such as severe loss of blood volume through hemorrhage, the net fluid reabsorption accompanying diminished intracapillary pressure helps to restore the volume of circulating fluid.

In the preceding discussion of transcapillary fluid balance, emphasis was placed on the difference in pressure between the capillaries and the interstitium rather than on the absolute values of the pressures in these compartments. In fact, there is still much controversy about what the actual pressures and protein concentrations are in interstitial fluid. It is commonly assumed that the interstitium has essentially zero hydrostatic pressure ($P_i \simeq 0$ mmHg) and negligible protein ($\pi_i \simeq 0$ mmHg), but the case for other values is equally strong. The controversy is about interstitial values, not about whether the Starling-Landis equation is correct.

Another complicating fact is that intracapillary pressure is not constant, but is higher at the entrance to a capillary than at the exit because of pressure losses due to resistance as the blood flows along capillaries. In fact, at the beginning of capillaries the capillary hydrostatic pressure exceeds the capillary oncotic pressure, whereas the reverse is true near the venous end of capillaries. Thus there is normally net fluid filtration in the beginning portions of capillaries and net fluid reabsorption in the final portions. A whole capillary, then, is in "net" equilibrium when its initial filtration and later reabsorption are equal. Fortunately, the net transcapillary fluid movement can be evaluated by using the average value of intracapillary pressure in the Starling-Landis equation.

LYMPHATIC SYSTEM

Protein molecules that have leaked into the interstitial tissue, as well as other large particles such as long-chain fatty acids or bacteria, cannot pass easily through capillary walls. If such particles were to accumulate in the interstitial space, filtration forces would ultimately exceed reabsorption forces and edema would result. The lymphatic system represents a pathway by which large molecules enter the circulating blood.

The lymphatic system begins in the tissues with blind-end lymphatic capillaries, which are roughly equivalent in size to but less numerous than regular capillaries. These capillaries are very porous and easily collect large particles accompanied by interstitial fluid. This fluid, called lymph, moves through the converging vessels; is filtered through lymph nodes, where bacteria are removed; and reenters the circulatory system near the point where the blood enters the right heart.

Flow of lymph from the tissues toward the entry point into the circulatory system is promoted by two factors: (1) increases in tissue interstitial pressure (due to fluid accumulation or to movement of surrounding tissue), and (2) contractions of the lymphatic vessels themselves. Valves located in these vessels also prevent backward flow.[5]

Roughly 2.5 liters of lymphatic fluid enters the cardiovascular system each day. When compared to the total amount of blood that circulates each day (about 7000 liters), this seems like an insignificant amount. However, lymphatic blockage is a very serious problem and is accompanied by severe swelling. Thus the lymphatics play a critical role in keeping the interstitial protein concentration low and in removing excess capillary filtrate from the tissues.

Study questions: **1** to **6**

[5] The question of how lymphatics collect fluid and protein and propel lymph toward the thorax is far from resolved. The interested reader should consult Zweifach and Silberberg (1979) for a more complete discussion.

Chapter 2

The Heart Pump

OBJECTIVES

The student understands the basic principles by which the heart pumps blood:

1. Identifies the chambers and valves of the heart and describes the pathway of blood flow through the heart.
2. Lists the factors essential to proper ventricular pumping action.

The student understands the ionic basis of the spontaneous electrical activity of cardiac muscle cells:

3. Describes how membrane potentials are created across semipermeable membranes by transmembrane ion concentration differences.
4. Defines equilibrium potential and knows its normal value for potassium and sodium ions.
5. States how membrane potential reflects a membrane's relative permeability to various ions.

6 Defines resting potential and action potential.
7 Defines threshold potential and describes the self-reinforcing interaction between ion permeability and membrane potential responsible for the depolarization phase of the action potential.
8 Defines pacemaker potential and describes the basis for rhythmic electrical activity of cardiac cells.
9 Lists the phase of the cardiac cell electrical cycle and states the membrane permeability alterations responsible for each phase.

The student knows the normal process of cardiac electrical excitation:

10 Describes gap junctions and their role in cardiac excitation.
11 Describes the normal pathway of action potential conduction through the heart.
12 Indicates the timing with which various areas of the heart are electrically excited, and identifies the characteristic action potential shapes and conduction velocities in each major part of the conduction system.

The student understands the physiological basis of the electrocardiogram:

13 States the relationship between electrical events of cardiac excitation and the P, QRS, and T waves, the PR interval, and the ST segment of the electrocardiogram.
14 States Einthoven's basic electrocardiographic conventions and, given data, determines the electrical axis of the heart.

The student knows the basic mechanical events of the cardiac cycle:

15 Defines and describes excitation-contraction coupling.
16 Lists the major distinct phases of the cardiac mechanical cycle as delineated by valve opening and closure.
17 Describes the pressure and volume changes in the atria, the ventricles, and the aorta during each phase of the cardiac cycle.
18 States the origin of the first and second heart sounds.
19 Defines and states normal values for (1) ventricular end-diastolic volume, end-systolic volume, stroke volume, diastolic pressure, and peak systolic pressure, and (2) aortic diastolic pressure, systolic pressure, and pulse pressure.
20 States similarities and differences between mechanical events in the left and right heart pump.

The student, through understanding normal cardiac function, can diagnose and appreciate the consequences of common cardiac abnormalities:

21 Detects common cardiac arrhythmias from the electrocardiogram and identifies their physiological basis.

22 Lists four common valvular abnormalities for the left heart and describes the alterations in heart sounds and intracardiac pressure and flow patterns that accompany them.

The sole function of the heart is to supply the energy required for the circulation of blood in the cardiovascular system. Blood flow through all organs is passive and occurs only because arterial pressure is kept higher than venous pressure by the pumping action of the heart. The right heart pump provides the energy necessary to move blood through the pulmonary vessels, and the left heart pump provides the energy that causes flow through the systemic organs.

Although the gross anatomy of the right heart pump is somewhat different from that of the left heart pump, the pumping principles are identical. Each pump consists of a ventricle, which is a closed chamber surrounded by a muscular wall, as illustrated in Fig. 2-1. The ventricle contains an outlet valve and an inlet valve. These valves are structurally designed to allow flow in only one direction and passively open and close in response to the direction of the pressure differences across them. The *pulmonic valve* is the outlet valve for the right ventricle, and the *aortic valve* is the outlet for the left ventricle. Preceding the inlet valve of each ventricle is another muscular heart chamber called an atrium. Consequently, the ventricu-

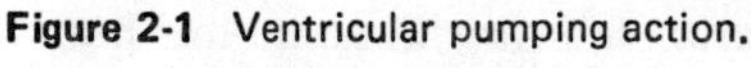

Figure 2-1 Ventricular pumping action.

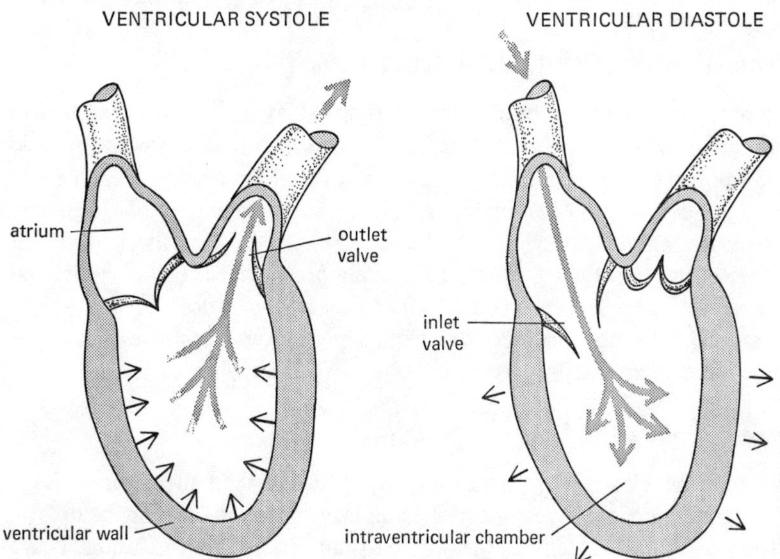

lar inlet valves are called the atrioventricular (AV) valves. The *tricuspid* is the AV valve for the right heart pump, and the *mitral* is the AV valve for the left heart pump.

Ventricular pumping action occurs because the volume of the intraventricular chamber is cyclically changed by rhythmic and synchronized contraction and relaxation of the individual cardiac muscle cells that form the ventricular wall. When ventricular muscle cells are contracting, blood is forced out of the ventricular chamber through the outlet valve, as shown in Fig. 2-1. This phase of the cardiac cycle is called *systole.* Because the pressure is higher in the ventricle than in the atrium during systole, the AV valve is closed. When the ventricular muscle cells relax, the pressure in the ventricle falls below that in the atrium, the AV valve opens, and the ventricle refills with blood. This portion of the cardiac cycle is called *diastole.* The outlet valve is closed during diastole because arterial pressure is greater than intraventricular pressure. After the period of diastolic filling, the systolic phase of a new cardiac cycle is initiated.

For effective, efficient ventricular pumping action, the heart must be functioning properly in four basic respects:

1 The contractions of individual cardiac muscle cells must occur at regular intervals and be synchronized (not *arrhythmic*).
2 The valves must open fully (not *stenotic*).
3 Closed valves must not leak (not *insufficient*).
4 The muscle contraction must be forceful (not *failing*).

We shall study in detail how these requirements are met in the normal heart.

ELECTRICAL ACTIVITY OF THE HEART

In all striated muscle cells, contraction is triggered by a rapid voltage change, called an *action potential*, that occurs on the cell membrane. Cardiac muscle cell action potentials differ sharply from those of skeletal muscle cells in three important ways that promote synchronous rhythmic excitation of the heart: (1) they can be self-generating; (2) they can be conducted directly from cell to cell; and (3) they have long durations, which preclude fusion of individual twitch contractions. To understand these special electrical properties of cardiac muscle and how cardiac function depends on them, we must first review the basic electrical properties of excitable cell membranes.

Membrane Potentials

All cells have an electrical potential (voltage) across their membranes. Such *membrane potentials* exist because the ionic concentrations of the cytoplasm are different from those of the interstitium and ions diffusing down concentration gradients across semipermeable membranes generate electrical gradients. The most important ions to consider are the potassium ion (K^+), which is more concentrated inside cells than in the interstitial fluid, and the sodium ion (Na^+), which has the opposite distribution.

Figure 2-2 shows how ion concentration differences can generate an electrical potential across the cell membrane. Consider first, as shown at the top of this figure, a cell that (1) has K^+ more concentrated inside the cell than out, (2) is permeable only to K^+, and (3) has no initial transmembrane potential. Because of the concentration difference, K^+ ions (positive charges) will diffuse out of the cell. Meanwhile, negative charges, such as protein anions, cannot leave the cell because the membrane is impermeable to them. Thus the K^+ efflux will make the inside of the cell more electrically negative (deficient in positively charged ions) and at the same time make the interstitium more electrically positive (rich in positive ions). Now K^+, being positively charged, is attracted to regions of electrical negativity. Therefore when K^+ diffuses out of a cell, it creates an electrical potential across the membrane that tends to attract it back into the cell. There exists one membrane potential called the *potassium equilibrium potential* at which the electrical forces tending to pull K^+ into the cell exactly balance the concentration forces tending to drive K^+ out. When the membrane potential has this value, there is no net movement of K^+ across the membrane. With the normal concentrations of about 145 m*M* K^+ inside cells and 4 m*M* K^+ in the extracellular fluid, the K^+ equilibrium potential is roughly −90 mV (inside more negative than outside by nine-hundreths of a volt).[1] A membrane that is permeable only to K^+ will inherently and rapidly (essentially instantaneously) develop the potassium equilibrium potential. In addition, membrane potential changes require the movement of so few ions that concentration differences are not significantly affected by the process.

As depicted in the bottom half of Fig. 2-2, similar reasoning shows how a membrane permeable only to Na^+ would have the *sodium equilibrium potential* across it. The sodium equilibrium potential is approximately +60 mV with the normal extracellular Na^+ concentration of 140 m*M* and intracellular concentration of 5 m*M*. Real cell membranes, however, are never permeable to just Na^+ or just K^+. When a membrane is permeable to both of these ions, the membrane potential will lie somewhere between the Na^+ equilibrium potential and the K^+ equilibrium potential. Just what membrane potential will exist any instant depends on the relative permeability of the

[1] The equilibrium potential (E_{eq}) for any ion (X^z) is determined by its intracellular and extracellular concentrations as indicated in the Nernst equation:

$$E_{eq}X^z = -z61.5 \log_{10} \frac{[X^z]_{inside}}{[X^z]_{outside}}$$

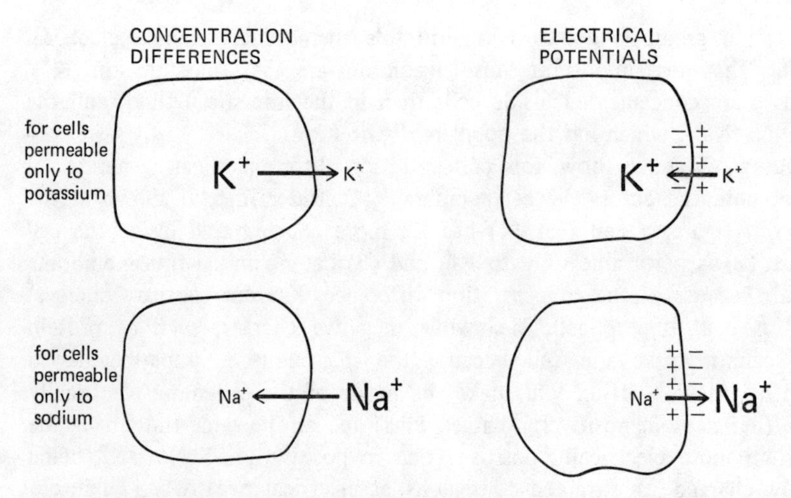

Figure 2-2 Electrochemical basis of membrane potentials.

membrane to Na^+ and K^+. The more permeable the membrane to K^+ than to Na^+, the closer the membrane potential will be to −90 mV. Conversely, when the permeability to Na^+ is high relative to the permeability to K^+, the membrane potential will be closer to +60 mV. The roles played by ions other than Na^+ and K^+ in determining membrane potential are usually minor and these ions may often be ignored. However, the calcium ion, Ca^{2+}, does participate in the cardiac muscle action potential. Like Na^+, Ca^{2+} is more concentrated outside cells than inside and the cell membrane tends to become more positive on the inside when the membrane's permeability to Ca^{2+} rises.

Under resting conditions, most heart muscle cells have membrane potentials that are quite close to the potassium equilibrium potential. Thus both electrical and concentration gradients favor Na^+ entry into the resting cell. However, the very low permeability of the resting membrane to Na^+ in combination with an energy-requiring sodium pump that extrudes Na^+ from the cell prevents Na^+ from gradually accumulating inside the resting cell.

Cardiac Muscle Action Potentials

Action potentials in all excitable cells are caused by abrupt, transient changes in the ionic permeability of the plasma membrane. Although the shape of the action potential differs somewhat between cardiac muscle cells and may change somewhat under various conditions, there are certain characteristics common to all cardiac cell action potentials. Figure 2-3 shows the form of a typical cardiac action potential along with the principal membrane permeability changes that cause it.

During the resting phase (phase 4), the membrane is more permeable to K^+ than to Na^+ so the membrane potential is low, i.e., near the potassium equilibrium potential. There is, however, a progressive decrease in the membrane's permeability to K^+ during the resting phase, which changes the Na^+-to-K^+ permeability ratio and causes a slow depolarization of the resting membrane. This spontaneous depolarization of the resting potential is alternatively called the *pacemaker potential*, diastolic depolarization, or phase 4 depolarization.

As shown in Fig. 2-3, the transition between the resting phase and the action potential phase is abrupt. The ratio of the membrane's permeability to Na^+ and K^+ suddenly reverses, and as a consequence the membrane potential shoots rapidly toward the sodium equilibrium potential. Although the molecular mechanisms behind these changes in membrane permeability are unknown, it is clear that the membrane's permeability properties somehow change with changes in membrane potential. Thus, ionic permeability controls membrane potential, and membrane potential in turn influences ionic permeability. This reciprocal interaction between membrane voltage and membrane permeability has the potential for being self-reinforcing, and this is precisely what happens during the rapidly rising phase (phase 0) of the action potential. Increased Na^+ permeability depolarizes the membrane, and this increases the Na^+ permeability still further. The influence of membrane

Figure 2-3 Time course of potential and permeability changes that occur during excitation.

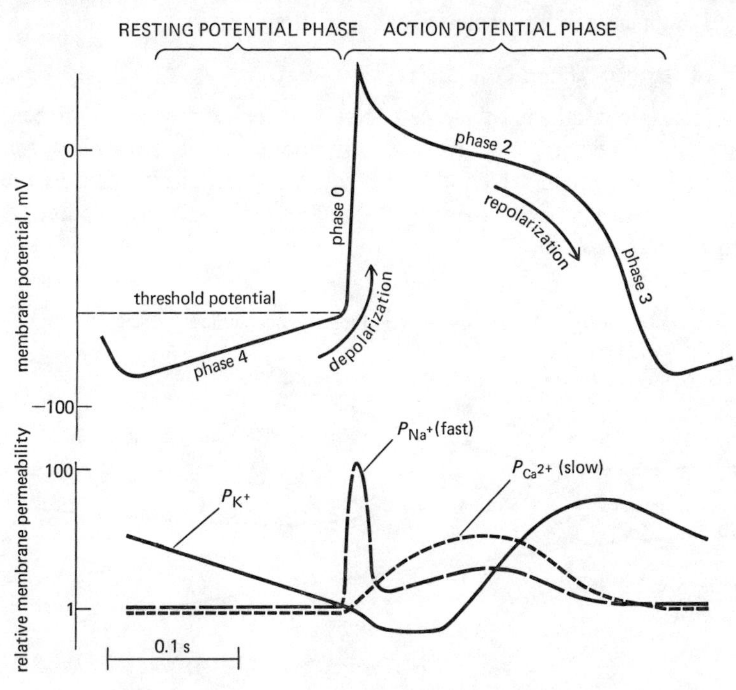

potential on Na^+ and K^+ permeabilities is such that self-reinforcing depolarization does not occur until the membrane voltage reaches a certain level called the *threshold potential.* Once the threshold potential is reached, as it ultimately will be by slow depolarization during the resting phase, an action potential will be automatically initiated by the self-regeneration phenomenon. An action potential, once initiated, cannot be stopped and proceeds to conclusion with permeability and voltage changes as shown in Fig. 2-3.

The sharp upswing (phase 0) of the action potential is caused primarily by the rapid increase in membrane Na^+ permeability. This initial abrupt permeability change (which produces what is sometimes referred to as the fast sodium current) is short-lived. It is followed by a more slowly developed increase in the membrane's permeability to Ca^{2+} and a decrease in its permeability to K^+. Also, there is a second and slowly developing increase in Na^+ permeability, which is thought to be caused by a different mechanism than those involved in the initial rapid Na^+ permeability changes. The more persistent permeability changes (which produce what is referred to as the slow inward current) prolong the depolarized state of the membrane to cause the plateau (phase 2) of the cardiac action potential. The membrane potential is repolarized (phase 3) to a resting level as the K^+ permeability increases and the Ca^{2+} and Na^+ permeabilities return to resting values. Differences in the characteristics of the resting and action potentials recorded in various areas of the heart primarily reflect regional differences in the magnitude and duration of the underlying permeability changes for these three ions.

Conduction of Cardiac Action Potentials

Action potentials are conducted from cell to cell in the heart because adjacent heart muscle cells have regions of close membrane association called *gap junctions* (nexuses) through which electrical currents can easily pass. Figure 2-4 shows schematically how these gap junctions allow action potential propagation from cell to cell.

Figure 2-4 Local currents and cell-to-cell conduction of cardiac muscle cell action potentials.

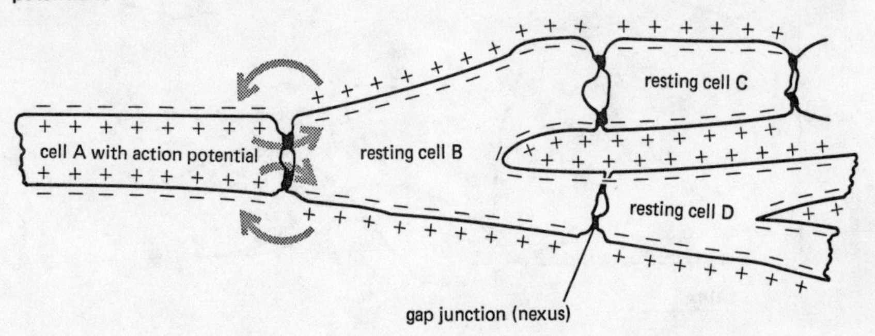

Cells B, C, and D are shown in the resting phase with more negative charges on the inside than the outside. Cell A is shown in the plateau phase of an action potential and has more positive charges inside than out. Because of the gap junctions, electrostatic attraction can cause a local current flow (ion movement) between the depolarized membrane of active cell A and the polarized membrane of resting cell B, as indicated by the arrows in the figure. This ion movement tends to eliminate the charge difference across the resting membrane; i.e., it depolarizes the membrane of cell B. Once the local currents from active cell A depolarize the membrane of cell B to the threshold level, an action potential will be triggered to cell B. This in turn will cause action potentials in C and D, and so on through the heart.

The speed at which an action potential propagates through a region of cardiac tissue is called the *conduction velocity.* The conduction velocity varies considerably in different areas in the heart. Two of the factors that favor a high conduction velocity are a large cell size and a steep depolarization phase of the action potential.

All the muscle cells of the heart possess inherent rhythmicity and are connected through gap junctions into a functional syncytium. However, there is a network of cardiac cells specifically adapted to generate the initial action potential for each heartbeat and conduct it throughout the heart. The major components of this specialized conduction system are shown in Fig. 2-5 and are the *sinoatrial node* (SA node), the *atrioventricular node* (AV node), and a ventricular conduction system composed of *Purkinje fibers.* Specific electrical adaptations of various cells in the heart are reflected in the characteristic shape of their action potentials, as shown in the right half of Fig. 2-5. Note that the action potentials shown in Fig. 2-5 have been positioned to indicate when the electrical impulse that originates in the SA node reaches other areas of the heart. Cells of the SA node act as the heart's normal pacemaker and determine the heart rate. This is because the spontaneous depolarization of the resting membrane is most rapid in SA nodal cells, and they reach their threshold potential before cells elsewhere.

The action potential initiated by an SA nodal cell spreads through the atrial wall in a wave centered at the SA node. Although there is some evidence for preferred conduction pathways in the atria, they are cetrainly not as well developed as the ventricle's Purkinje system. Action potentials from cells in two different regions of the atria are shown in Fig. 2-5: one close to the SA node and one more distant from the SA node. Both cells have similarly shaped action potentials, but their temporal displacement indicates that it takes some time for the impulse to spread over the atria. The atrial conduction velocity is about 1 m/s, and the impulse reaches the AV node roughly 0.08 s after having been generated in the SA node.

The AV node forms the only bridge of contiguous cardiac cells crossing the cartilaginous structure that separates atria from ventricles. As shown in

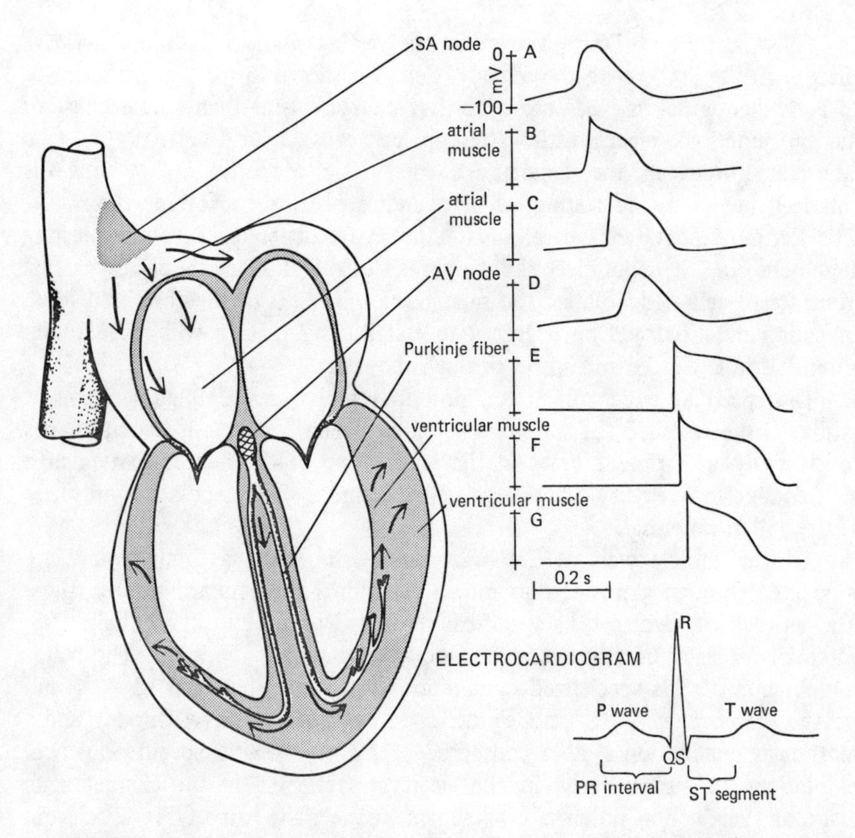

Figure 2-5 Electrical activity of the heart: single-cell voltage recordings (traces A to G), and lead II electrocardiogram.

Fig. 2-5, cells of the AV node have action potentials similar in shape to those of SA nodal cells. Because the cardiac impulse travels very slowly through the AV nodal tissue (0.05 m/s), it takes the impulse about 0.15 s to cross the short AV nodal region. Since the AV node delays the transfer of the cardiac impulse from the atria to the ventricles, the ventricles contract slightly after the atria in each cardiac cycle. Note also that AV nodal cells have a faster spontaneous resting depolarization than other cells of the heart except those in the SA node. The AV node is sometimes referred to as a *latent pacemaker*, and in many pathological situations it, rather than the SA node, controls the heart rhythm.

The cardiac impulse normally exits the AV node into a bundle of ventricular Purkinje fibers called the *bundle of His.* This "common bundle" soon divides into the *left bundle branch* and the *right bundle branch*, which run down either side of the intraventricular septum. After multiple branching into finer and finer pathways, Purkinje fibers ultimately terminate on ordinary ventricular muscle cells in many areas of the ventricle. Because of sharply rising action potentials and other factors, such as large cell diameters, electrical conduction is extremely rapid in Purkinje fibers (~3 m/s). This allows the Purkinje system to transfer the cardiac impulse to cells in many areas of the ventricle nearly in unison.

Action potentials from muscle cells in two areas of the ventricle are shown in Fig. 2-5. Because of the high conduction velocity, there is only a small discrepancy in their time of onset.

Electrocardiogram

Fields of electrical potential caused by the electrical activity of the heart extend through the body tissue and may be measured with electrodes placed on the body surface. *Electrocardiograms* are records of how the voltage between two points on the body surface changes with time as a result of the electrical events of the cardiac cycle. At any instant of the cardiac cycle the electrocardiogram indicates the net electrical field that is the summation of many weak electrical fields being produced by individual cardiac cells at that instant. When a large number of cells are simultaneously depolarizing or repolarizing, large voltages are observed on the electrocardiogram. Since the electrical impulse spreads through the heart tissue in a stereotyped manner, the temporal pattern of voltage change recorded between two points on the body surface is also stereotyped and will repeat itself with each heart cycle.

The lower trace of Fig. 2-5 represents a typical recording of the voltage changes normally measured between the right arm and the left leg as the heart goes through one cycle of electrical excitation; this record is called a lead II electrocardiogram. The major features of an electrocardiogram are the *P wave*, the *QRS complex*, and the *T wave*. The P wave corresponds to atrial depolarization, the QRS complex to ventricular depolarization, and the T wave to ventricular repolarization.

The period of time from the initiation of the P wave to the beginning of the QRS complex is designated as the PR interval and indicates the time it takes for an action potential to spread through the atria and the AV node. During the later portion of the PR interval no voltages are detected on the body surface. This is because atrial muscle cells are all depolarized (in the plateau phase of their action potentials), ventricular cells are still resting, and the electrical field set up by the action potential progressing through the small AV node is not intense enough to be detected. Shortly after the cardiac impulse breaks out of the AV node and into the rapidly conducting Purkinje system, all the ventricular muscle cells depolarize within a very short period of time and cause the QRS complex. The R wave is the largest event in the electrocardiogram because ventricular muscle cells are so numerous and because they depolarize nearly in unison.

The QRS complex is followed by a period called the *ST segment.* Normally, no electrical potentials are measured on the body surface during the ST segment because no rapid changes in membrane potential are occurring in any of the cells of the heart; atrial cells have already returned to the resting phase, whereas ventricular muscle cells are in the plateau phase of their action potentials. Once ventricular cells begin to repolarize, however, a voltage once again appears on the body surface and is measured as the T wave of the electrocardiogram. The T wave is broader and not as large as the R wave because ventricular repolarization is less synchronous than depolarization. At the conclusion of the T wave all the cells in the heart are in the resting state. No body surface potential is measured until the next impulse is generated by the SA node.

It should be recognized that the operation of the specialized conduction system is a primary factor in determining the normal electrocardiographic pattern. For example, the AV nodal transmission time determines the PR interval. Also, the effectiveness of the Purkinje system in synchronizing ventricular depolarization is reflected in the large magnitude and short duration of the R wave. It should also be recalled that every heart muscle cell is inherently capable of rhythmicity and that all cardiac cells are electrically interconnected through gap junctions. Thus a functional heart rhythm can and often does occur without the involvement of part or all of the specialized conduction system. Such a situation is, however, abnormal, and the existence of abnormal conduction pathways would lead to an abnormal electrocardiogram.

Basic Electrocardiographic Conventions Recording electrocardiograms has become a routine diagnostic procedure, which has been standardized by universal application of certain conventions. The conventions for recording and analysis of electrocardiograms from the three standard limb leads are briefly described here.

Recording electrodes are placed on both arms and the left leg—usually at the wrists and ankle. The assumptions are made that the appendages act merely as extensions of the recording system and that voltage measurements are made between points that form an equilateral triangle over the thorax, as shown in Fig. 2-6. This conceptualization is called Einthoven's triangle in honor of the Dutch physiologist who devised it at the turn of the century. Any single electrocardiographic trace is a recording of the voltage difference measured between any two vertices of Einthoven's triangle. We have already discussed the lead II electrocardiogram measured between the right arm and the left leg. Similarly, lead I and lead III electrocardiograms represent voltage measurements taken along the other two sides of Einthoven's triangle, as indicated in Fig. 2-6. The + and − symbols in Fig. 2-6 indicate polarity conventions that have been universally adopted. For example, an upward

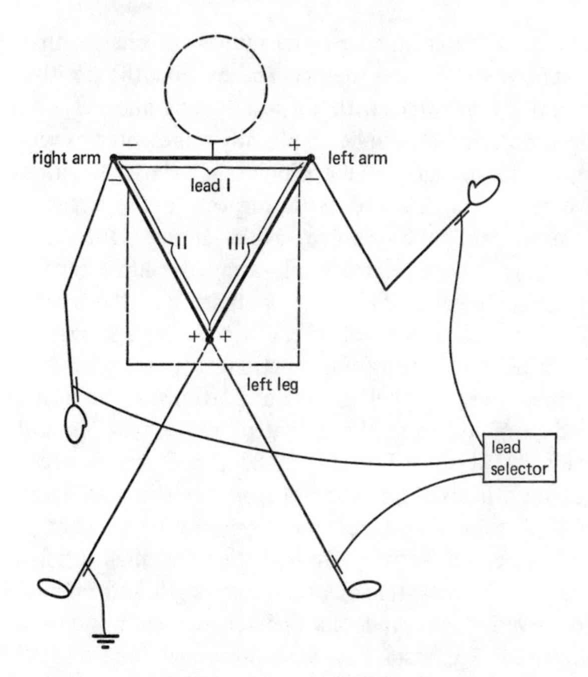

Figure 2-6 Einthoven's electrocardiographic conventions.

deflection in a lead II electrocardiogram (as normally occurs during the P, R, and T waves) indicates that the voltage measured at the left leg is more positive than that at the right shoulder. Similar polarity conventions have been established for lead I and lead III recordings and are indicated by the + and − symbols in Fig. 2-6. In addition, electrocardiographic recording equipment has been standardized so that 1 cm on the vertical axis always represents a potential difference of 1 mV, and 1 s is represented by 25 mm on the horizontal axis of any electrocardiographic record.

As shown later in this chapter, many cardiac electrical abnormalities can be detected in recordings from a single electrocardiographic lead. However, certain clinically useful information must be derived by combining the information obtained from two electrocardiographic leads. To understand these more complex electrocardiographic analyses, we must first examine more closely how voltages appear on the body surface as a result of the cardiac electrical activity.

Cardiac Dipoles and the Electrical Axis At any instant in the cardiac cycle the electrical fields being generated by the heart may be viewed as originating from a simple electrical dipole (a pair of spatially distinct regions, one of which is positively charged and one of which is equally but negatively charged). A dipole has a strength (determined by the number of charges that were separated to produce it) and a direction (determined by how the positive and negative regions are spatially oriented with respect to one another). As shown in Fig. 2-7A to C, electrical dipoles can be graphically represented as vectors whose length represents the strength of the dipole and whose direction indicates how the positive region is oriented with respect to the negative region. Thus at every instant of the cardiac cycle–during ventricular depolarization at the peak of the R wave, for example–the heart has a certain dipole strength and a certain dipole orientation. The orientation of the heart's dipole at the instant the R wave reaches its peak is called the *electrical axis of the heart.* It is clinically useful in detecting nonfunctioning areas of cardiac muscle (infarcts) and abnormal growth (hypertrophy) of portions of the heart. The electrical axis can be determined from the height of the R wave in any two of the three limb leads, as illustrated in Fig. 2-8. The R waves occur simultaneously in the three limb leads since all stem from the common event of ventricular depolarization. The limb lead configuration may be thought of as a way to view the heart's electrical activity from three different positions. The vector representing the heart's instantaneous dipole strength and orientation is the object under observation, and it looks different depending on the position from which it is viewed. The instantaneous voltage measured on lead I, for example, indicates how the dipole being generated by the heart's electrical activity at that instant appears when viewed from directly above. A cardiac dipole that is oriented horizontally appears large on lead I, whereas a vertically oriented cardiac dipole, however large, is not detectable on lead I. Thus it is necessary to have views from two directions to establish the magnitude and orientation of the heart's dipole. A vertically oriented dipole would be invisible on lead I but would be readily apparent if viewed from the perspective of lead II or lead III.

The magnitude and orientation of the heart's electrical dipole change continually throughout the cardiac cycle, and this produces the variations in body surface potential measured in the limb leads. The cardiac dipole vector shown in the

Figure 2-7 Electrical dipoles and their representation as vectors.

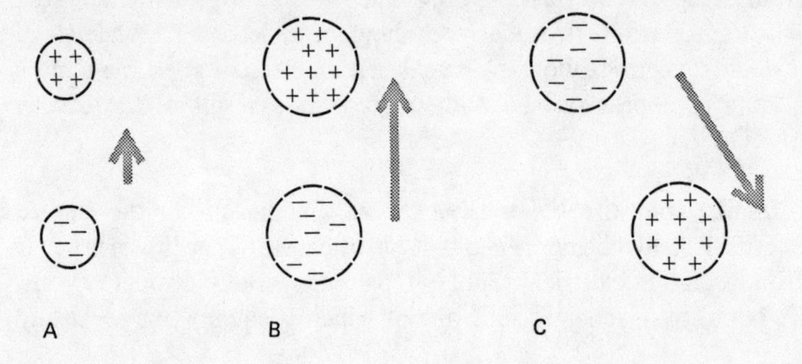

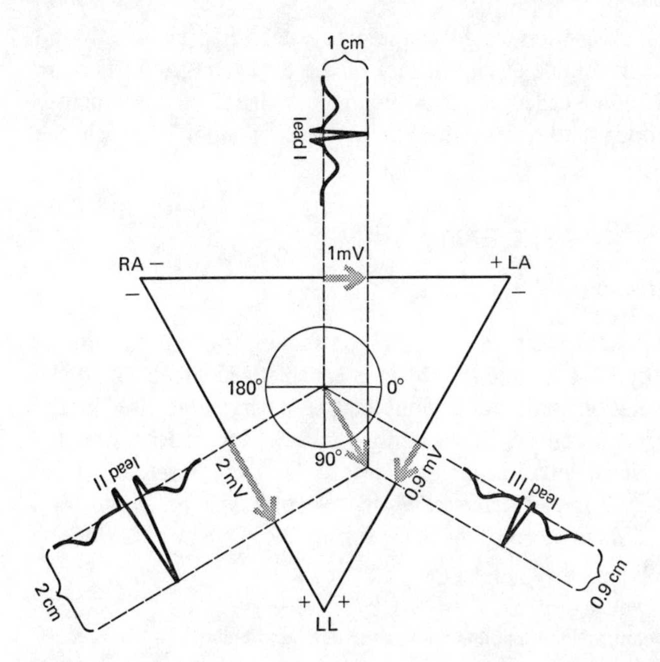

Figure 2-8 Use of limb lead information to determine the electrical axis of the heart.

center of Fig. 2-8 represents the magnitude and direction of the heart's dipole as if its continuous fluctuations were frozen at the height of ventricular depolarization. Its magnitude and orientation were deduced from the lead I and lead II recordings in the following manner. In lead I, peak R wave deflection of 1 cm upward indicates that the horizontal component of the heart's dipole at this instant has a strength of 1 mV. The upward deflection on lead I indicates that the heart's dipole points toward the left shoulder when viewed from above. Similarly, the R wave in lead II indicates that the component of the heart's dipole projected on the lead II axis has a magnitude of 2 mV and is directed toward the left leg. Inspection of Fig. 2-8 reveals that the cardiac dipole vector shown is the only one that could simultaneously produce the lead I and lead II observations. It may be formally obtained by orthographically projecting the lead I and lead II components to the center of Einthoven's triangle. The axis of the heart in this example is +59°. By convention, this angle is measured from the horizontal as indicated in Fig. 2-8. An electrical axis lying anywhere in the lower right-hand quadrant of the graph (0° → +90° in the electrocardiographic terms) is considered normal. Note that the use of leads I and II to determine the electrical axis of the heart is an arbitrary choice; the electrical axis can be determined from the information in any two leads. Also keep in mind that the electrical axis of the heart represents the orientation of the cardiac dipole during only one instant in the cardiac cycle. Similar procedures could be used to establish the strength and orientation of the heart's dipole at any instant in the cardiac cycle. In fact, an advanced analysis technique called vectorcardiography is based on continuously following the magnitude and orientation of the heart's dipole through the cardiac cycle.

MECHANICAL EVENTS OF THE HEART

Cardiac Muscle Contraction

Muscle action potentials trigger mechanical contraction through a process called *excitation-contraction coupling*, which is schematized in Fig. 2-9. The major event of excitation-contraction coupling is a dramatic rise in the intracellular free Ca^{2+} concentration. In cardiac muscle, two mechanisms are responsible for the rise in intracellular Ca^{2+} levels: (1) Ca^{2+} is released from an intracellular storage compartment called the *sarcoplasmic reticulum*, and (2) extracellular Ca^{2+} diffuses into the cell during the plateau phase of the cardiac action potential.

Figure 2-9 Excitation-contraction coupling and sarcomere shortening. A. Resting. B. Contracting.

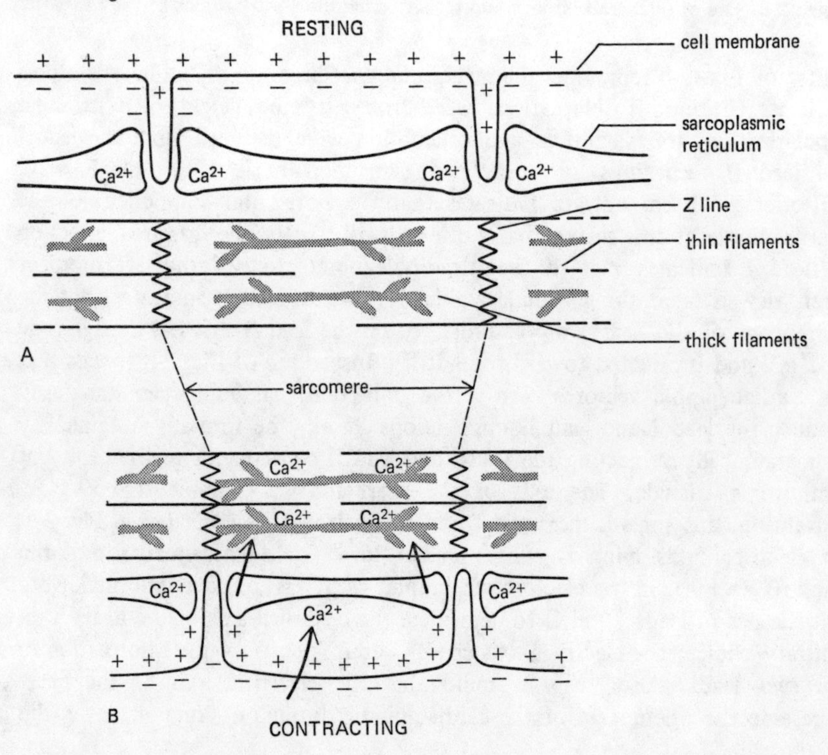

When the intracellular Ca^{2+} level is high, links called *cross bridges* form between two sets of filaments, or myofilaments, found within muscle. As indicated in Fig. 2-9, *thick filaments*, composed of the contractile protein *myosin*, and *thin filaments*, composed of other contractile proteins including *actin* and a regulatory protein *troponin*, are arranged in a regular and parallel manner within the basic contractile unit of muscle, which is called a *sarcomere.* Sarcomere units, as depicted in Fig. 2-9, are joined end to end at Z lines to form *myofibrils*, which run the length of the muscle cell. During contraction, thick and thin filaments slide past one another to shorten each sarcomere and thus the muscle as a whole. The forces that cause muscle force development and/or shortening are somehow generated by the Ca^{2+}-induced formation of cross bridges between the thick and thin myofilaments. The bridges form when regularly spaced myosin protrusions from thick filaments attach to regularly spaced sites on the actin molecules in the thin filaments. This actin-myosin interaction requires energy from adenosine triphosphate (ATP). In resting muscles, the attachment of myosin to the actin sites is inhibited by troponin. Calcium causes muscle contraction by interacting with troponin to remove its inhibition of the actin sites. Since a single cross bridge is a very short structure, gross muscle shortening requires that cross bridges repetitively form, produce incremental movement between the myofilaments, detach, form again at a new actin site, and so on in a cyclic manner.

Contraction presumably terminates when the Ca^{2+} concentration around the myofilaments decreases. The mechanisms by which this is accomplished are not fully understood, but active sequestration of Ca^{2+} into the sarcoplasmic reticulum is undoubtedly involved. Excitation-contraction coupling in cardiac muscle is different from that in skeletal muscle in that it may be modulated; different intensities of actin-myosin interaction (contraction) can result from a single action potential trigger in cardiac muscle. The mechanism for this seems to be variations in the amount of Ca^{2+} reaching the myofilaments during the twitch. This ability of cardiac muscle to vary its contractile strength–i.e., change its *contractility*–is extremely important to cardiac function, as discussed in Chap. 3.

Cardiac Cycle—Left Pump

The mechanical function of the heart is reflected in the pressure, volume, and flow changes that occur within it during the cardiac cycle. The normal mechanical events of a cycle of the left heart pump are correlated in Fig. 2-10. This important figure summarizes a great deal of information and should be studied carefully.

Ventricular Diastole The *diastolic phase*[2] of the cardiac cycle begins with the opening of the AV valves. As shown in Fig. 2-10, the mitral valve opens

[2] The atria and ventricles do not beat simultaneously. Usually, and unless otherwise noted, systole and diastole denote phases of ventricular operation.

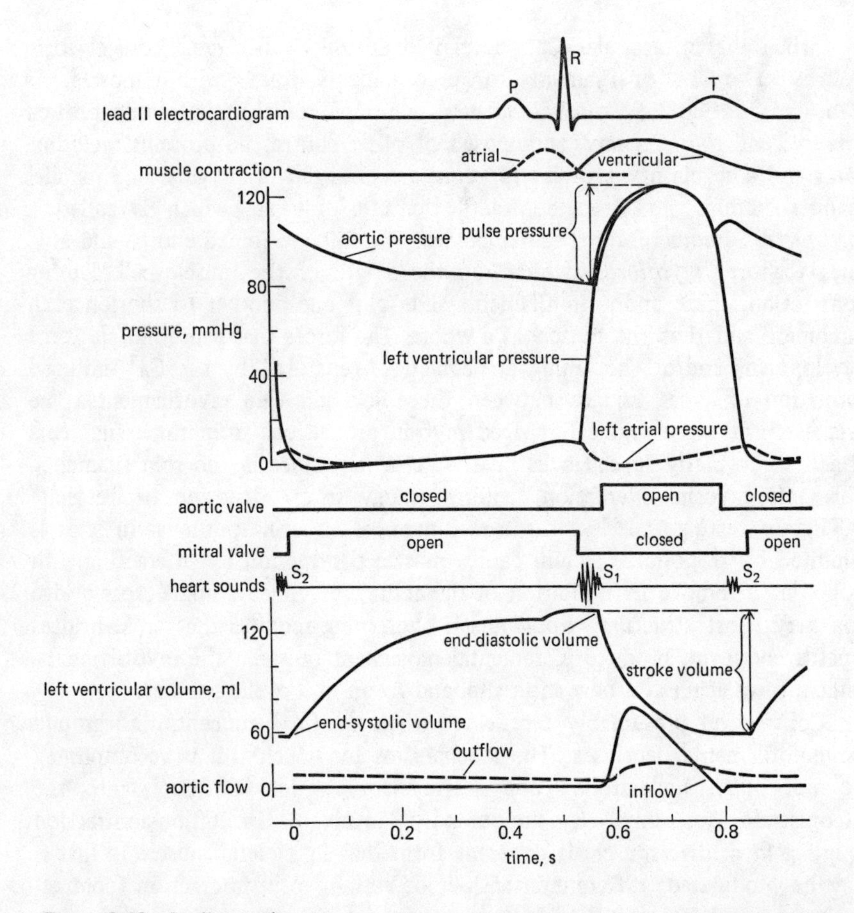

Figure 2-10 Cardiac cycle.

when left ventricular pressure falls below left atrial pressure and the period of ventricle filling begins. Blood that had previously accumulated in the atria behind the closed mitral valve empties rapidly into the ventricle and this causes an initial drop in atrial pressure. Later, the pressures in both chambers slowly rise together as the atria and ventricle continue filling in unison with blood returning to the heart through the veins.

Atrial contraction is initiated near the end of ventricular diastole by the depolarization of the atrial muscle cells, which causes the P wave of the electrocardiogram. As the atrial muscle cells develop tension and shorten, atrial pressure rises and an additional amount of blood is forced into the ventricle. At normal heart rates, atrial contraction is not essential for adequate ventricular filling. This is evident in Fig. 2-10 from the fact that the ventricle has nearly reached its maximum or *end-diastolic volume* before atrial contraction begins. Atrial contraction plays an increasingly significant role in ventricular filling as heart rate increases because the time interval between beats for passive filling becomes progressively shorter with increased heart rate. Note that throughout diastole, atrial and ventricular pressures are nearly identical. This is because a normal open mitral valve has very little resistance to flow and thus only a very small atrial-ventricular pressure difference is necessary to produce ventricular filling.

Ventricular Systole Ventricular systole begins when the action potential breaks through the AV node and sweeps over the ventricular muscle–an event heralded by the QRS complex of the electrocardiogram. Contraction of the ventricular muscle cells causes intraventricular pressure to rise above that in the atrium, which causes abrupt closure of the AV valve. Mechanical vibrations caused by AV valve closure are audible on the chest surface as the *first heart sound.*

Pressure in the left ventricle continues to rise sharply as the ventricular contraction intensifies. When the left ventricular pressure exceeds that in the aorta, the aortic valve opens. The period of time between mitral valve closure and aortic valve opening is referred to as the *isovolumetric contraction phase* because during this interval the ventricle is a closed chamber with a fixed volume. Ventricular ejection begins with the opening of the aortic valve. In early ejection blood enters the aorta rapidly and causes the pressure there to rise. Pressure builds simultaneously in both the ventricle and the aorta as the ventricular muscle cells continue to contract in early systole. This period is often called the *rapid ejection phase.*

Left ventricular and aortic pressure ultimately reach a maximum called *peak systolic pressure.* At this point the strength of ventricular muscle contraction begins to wane. Muscle shortening and ejection continue, but at a reduced rate. Aortic pressure begins to fall because blood is leaving the aorta and large arteries faster than blood is entering from the left ventricle. Throughout ejection, very small pressure differences exist between the left ventricle and the aorta because the aortic valve orifice is so large that it presents very little resistance to flow.

Eventually, the strength of the ventricular contraction diminishes to the point where intraventricular pressure falls below aortic pressure. This causes abrupt closure of the aortic valve and mechanical vibrations audible as the *second heart sound.* A dip, called the incisura or *dicrotic notch*, appears in the aortic pressure trace because a small volume of aortic blood must flow backward to fill the aortic valve leaflets as they close. Note that throughout ventricular systole, blood continues to return to the heart and fill the atria. Thus atrial pressure progressively rises during ventricular systole to the high value that promotes rapid ventricular filling once the AV valve opens to begin the next heart cycle.

After aortic valve closure, intraventricular pressure falls rapidly as the ventricular muscle relaxes. For a brief interval, called the *isovolumetric relaxation period*, the mitral valve is also closed. Ultimately, intraventricular pressure falls below atrial pressure, the AV valve opens, and a new cardiac cycle begins. The ventricle has reached its minimum or *end-systolic volume* at the time of aortic valve closure. The amount of blood ejected from the ventricle during a single beat, the *stroke volume*, is equal to ventricular end-diastolic volume minus ventricular end-systolic volume.

The aorta distends or balloons out during systole because more blood enters the aorta than leaves it. During diastole, the arterial pressure is maintained by the elastic recoil of walls of the aorta and other large arteries. Nonetheless, aortic pressure gradually falls during diastole as the aorta supplies blood to the systemic vascular beds. The lowest aortic pressure, reached at the end of diastole, is called *diastolic pressure*. The difference between diastolic and peak systolic pressure in the aorta is called the arterial *pulse pressure*. Typical values for systolic and diastolic pressures in the aorta are 120 and 80 mmHg, respectively.

Cardiac Cycle—Right Pump

Because the entire heart is served by a single electrical excitation system, similar mechanical events occur essentially simultaneously in both the left heart and the right heart. Both ventricles have synchronous systolic and diastolic periods and the valves of the right and left heart normally open and close nearly in unison. Since the two sides of the heart are arranged in series in the circulation, they must pump the same amount of blood and therefore must have identical stroke volumes.

The major difference between the right and left pumps is in the magnitude of the peak systolic pressure. The lungs provide considerably less resistance to blood flow than that offered collectively by the systemic organs. Therefore less arterial pressure is required to drive the cardiac output through the lungs than through the systemic organs. Typical pulmonary artery systolic and diastolic pressures are 24 and 8 mmHg, respectively.

PUMP ABNORMALITIES

Sometimes the heart does not function properly because individual muscle cells fail to contract forcefully enough. This situation is called *cardiac failure* and will be discussed in Chap. 8. Functional abnormalities may also occur either in the heart's electrical excitation process or in its valves. The diagnosis and functional significance of some common abnormalities of these latter types are described briefly in the following two sections.

Electrical Abnormalities and Arrhythmias

Many cardiac excitation problems can be diagnosed from the information in a single lead of an electrocardiogram, as illustrated in Fig. 2-11.

Paroxysmal atrial tachycardia results when a highly irritated (rapidly firing) cell somewhere in the atrium takes over the pacemaker function and drives the heart at a very high rate. A region outside the SA node that originates cardiac impulses is called an *ectopic focus.* QRS complexes appear normal (albeit frequent) with simple paroxysmal atrial tachycardia because the ventricular conduction pathways operate normally. The P and T waves may be superimposed because of the high heart rate. Low blood pressure and fainting may accompany bouts of this arrhythmia because the extremely high heart rate does not allow sufficient diastolic time for ventricular filling.

In *first-degree heart block* the only electrical abnormality is unusually slow conduction through the AV node. This condition is detected by an abnormally long PR interval (usually >0.2 s). Otherwise, the electrocardiogram may be completely normal. At normal heart rates the physiological effects of a first-degree block are inconsequential.

A *second-degree heart block* is said to exist when some but not all atrial

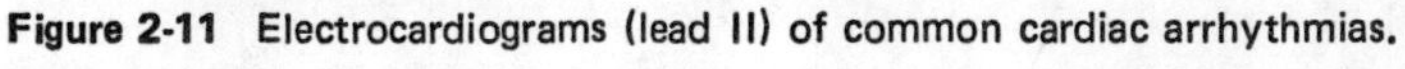

Figure 2-11 Electrocardiograms (lead II) of common cardiac arrhythmias.

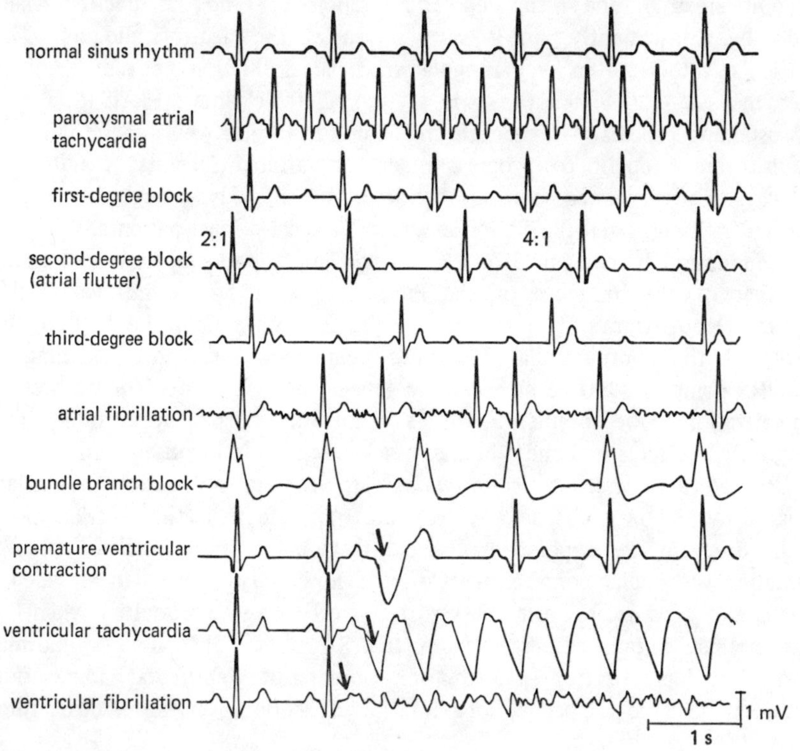

impulses are transmitted through the AV node to the ventricle. Impulses are blocked in the AV node if the cells of the region are still in a refractory period from a previous excitation. The situation is aggravated by high atrial rates and slower than normal conduction through the AV nodal region. In second-degree block, some but not all P waves are accompanied by corresponding QRS complexes and T waves. Atrial rate is often faster than ventricular rate by a certain ratio (e.g., 2:1, 3:1, 4:1). This condition may not represent a serious clinical problem as long as the ventricular rate is adequate to meet the pumping needs. The term *atrial flutter* is often applied when very high atrial rates occur and are not accompanied by high ventricular rates.

In *third-degree heart block*, no impulses are transmitted through the AV node. Some area in the ventricles–often in the common bundle or bundle branches near the exit of the AV node–assumes the pacemaker role for the ventricular tissue. Atrial rate and ventricular rate are completely independent, and P waves and QRS complexes are totally dissociated in the electrocardiogram. Ventricular rate is likely to be slower than normal (bradycardia) and sometimes is slow enough to impair cardiac output.

Atrial fibrillation is characterized by a complete loss of the normally close synchrony of the excitation and resting phases between atrial cells. Cells in different areas of the atria depolarize, repolarize, and are excited again randomly. Consequently, no P waves appear in the electrocardiogram. The ventricular rate is often very irregular in atrial fibrillation because impulses enter the AV node from the atria at unpredictable times. Fibrillation is a self-sustaining process. The mechanisms behind it are not well understood, but impulses are thought to progress repeatedly around irregular conduction pathways. However, because atrial contraction usually plays a negligible role in ventricular filling, atrial fibrillation is well tolerated by most patients.

Conduction blocks called *bundle branch blocks* or *hemiblocks* can occur in either of the branches of the Purkinje system of the intraventricular septum. Depolarization is less synchronous than normal in the half of the heart with the nonfunctional Purkinje system. This results in a widening of the QRS complex (>0.12 s) because a longer time is required for ventricular depolarization to be completed (0.12 s is the usual normal upper limit). The physiological effects of bundle branch blocks are usually inconsequential.

Premature ventricular contractions are caused by action potentials initiated by and propagated away from an ectopic focus in the ventricle. As a result, the ventricle depolarizes and contracts before it normally would. A premature ventricular contraction is often followed by a missed beat called a *compensatory pause*, because the ventricular cells are still refractory when the next normal impulse emerges from the SA node. The highly abnormal ventricular depolarization pattern of a premature ventricular contraction produces the large-amplitude, long-duration deflections on the electrocardiogram. The shapes of the electrocardiographic records of these extra beats are highly variable and depend on the ectopic site of their origin and the depolarization pathways involved. The volume of blood ejected by the premature beat iself is smaller than normal, whereas the stroke volume of the beat following the compensatory pause is larger than normal. This is due partly to the differences in filling times and partly to an inherent phenomenon of cardiac muscle called postextrasystolic potentiation. Single premature ventricular contractions occur occasionally in most individuals and are not dangerous.

Ventricular tachycardia occurs when the ventricles are driven at high rates by impulses originating from ventricular ectopic foci. Ventricular tachycardia is a very serious condition. Not only is diastolic filling time limited by the rapid rate, but the abnormal excitation pathways make ventricular contraction less synchronous and therefore less effective than normal. In addition, ventricular tachycardia often precedes ventricular fibrillation.

In *ventricular fibrillation*, various areas of the ventricle are excited and contract asynchronously. The mechanisms are similar to those in atrial fibrillation. Since no pumping action occurs with ventricular fibrillation, the situation is fatal unless quickly corrected by cardiac conversion. During conversion, the artificial application of large currents to the entire heart may be effective in depolarizing all heart cells simultaneously and thus allowing a normal excitation pathway to be reestablished.

Valvular Abnormalities

Pumping action of the heart is also impaired when the valves do not function properly, and abnormal heart sounds usually accompany cardiac valvular defects. The abnormal sounds, called *murmurs*, are caused by abnormal pressure gradients and blood flow patterns that occur during the cardiac cycle. A number of techniques, ranging from simple auscultation (listening to the heart sounds) to cardiac catheterization, are used to obtain information about the nature and extent of the malfunction. A brief overview of four of the common valve defects is given in Fig. 2-12.

Aortic Stenosis Normally, the aortic valve represents a pathway of very low resistance through which blood leaves the left ventricle. If this opening is narrowed (stenotic), its resistance increases. A significant pressure difference between the left ventricle and the aorta may be required to eject blood through a stenotic aortic valve. As shown in Fig. 2-12, intraventricular pressures may rise to very high levels during systole while aortic pressure rises more slowly than normal to a systolic value that is subnormal. Pulse pressure is usually low with aortic stenosis. High intraventricular pressure development is a strong stimulus for cardiac muscle cell hypertrophy, and an increase in

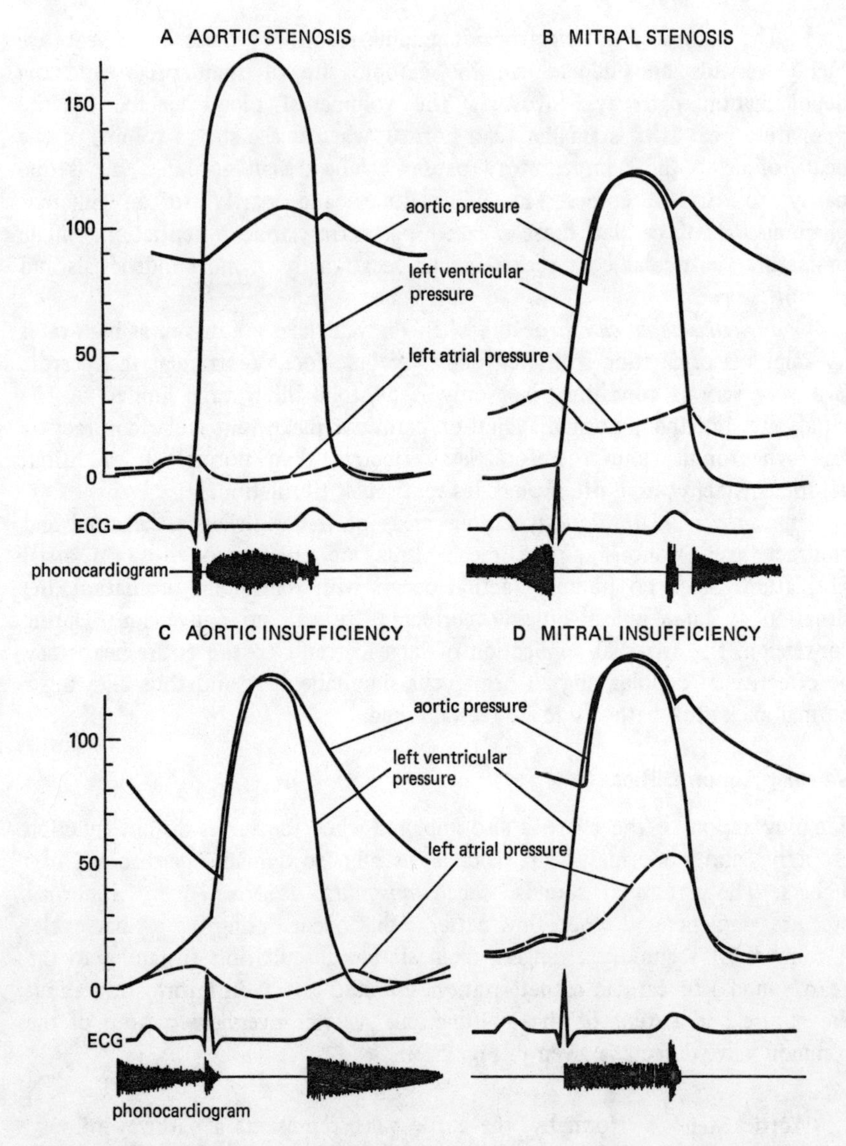

Figure 2-12 Common valvular abnormalities. A. Aortic stenosis. B. Mitral stenosis. C. Aortic insufficiency. D. Mitral insufficiency.

left ventricular muscle mass invariably accompanies aortic stenosis. This tends to produce a leftward deviation of the electrical axis. (The mean electrical axis will fall in the upper right-hand quadrant of the graph in Fig. 2-8.) Blood being ejected through the narrowed orifice may reach very high velocities and turbulent flow may occur in the aorta. This abnormal turbulent flow can be heard as a *systolic* (or ejection) *murmur* with a properly placed stethoscope.

Mitral Stenosis A pressure difference of more than a few millimeters of mercury across the mitral valve during diastole is distinctly abnormal and indicates that this valve is stenotic. The high resistance mandates an elevated pressure difference to achieve normal flow across the valve ($\dot{Q} = \Delta P/R$). Consequently, left atrial pressure and volume are elevated with mitral stenosis. A diastolic murmur associated with turbulent flow through the stenotic valve can often be heard.

Aortic Regurgitation (Insufficiency) When the leaflets of the aortic valve do not provide an adequate seal, blood regurgitates from the aorta back into the left ventricle during the diastolic period. As shown in Fig. 2-12, aortic pressure falls faster and further than normal during diastole, which causes a low diastolic pressure and a large pulse pressure. In addition, ventricular end-diastolic volume and pressure are higher than normal because of the extra blood that reenters the chamber through the incompetent aortic valve during diastole. Turbulent flow of the blood reentering the left ventricle during early diastole produces a characteristic *diastolic murmur.* Often the aortic valve is altered so that it is both stenotic and insufficient. In these instances both a systolic and a diastolic murmur are present.

Mitral Insufficiency When the mitral valve is insufficient some blood regurgitates from the left ventricle into the left atrium during systole. A systolic murmur may accompany this abnormal flow pattern. As shown in Fig. 2-12, left atrial pressure is raised to abnormally high levels, and left ventricular end diastolic volume and pressure increase.

Study questions: **7** to **14**

Chapter 3

Determinants and Control of Cardiac Output

OBJECTIVES

The student understands the factors that determine cardiac output:

1 States the relationship between cardiac output, heart rate, and stroke volume.
2 States how diastolic potentials of pacemaker cells can be altered to produce changes in heart rate.
3 Describes how cardiac sympathetic and parasympathetic nerves alter heart rate.
4 States Starling's law of the heart.
5 States the law of Laplace.
6 Describes the ventricular volume-pressure cycle and the cardiac muscle length-tension cycle and the correlation between them.
7 Describes the active and passive length-tension relationships for cardiac muscle.
8 Defines isometric, isotonic, and afterloaded contractions of cardiac muscle.

9 Describes the influence of altered preload on the tension-producing and shortening capabilities of cardiac muscle.
10 Describes the influence of altered afterload on the shortening capabilities of cardiac muscle.
11 Defines ventricular preload and ventricular afterload.
12 Predicts the effects of altered ventricular preload and afterload on ventricular stroke volume.
13 Defines cardiac contractility.
14 Describes how changes in contractility alter stroke volume.
15 Describes the effect of cardiac sympathetic nerves on contractility, stroke volume, and cardiac output.
16 Draws a family of cardiac function curves describing the relationship between filling pressure and cardiac output under various levels of sympathetic tone.
17 Given data, calculates cardiac output using the Fick principle.

Cardiac output (liters of blood pumped by *each* of the ventricles per minute) is an extremely important cardiovascular variable that is continuously adjusted so that the cardiovascular system operates to meet the body's moment-to-moment transport needs. In going from rest to strenuous exercise, for example, the cardiac output of an average person will increase from approximately 5.8 to perhaps 15 liters/min. The extra cardiac output provides the exercising skeletal muscles with the additional nutritional supply needed to sustain an increased metabolic rate. To understand the cardiovascular system's response not only to exercise but to all other physiological or pathological demands placed on it, we must understand what determines and controls cardiac output.

Cardiac output (CO) is determined by the amount of blood ejected with each beat (SV) and the number of heartbeats per minute (HR) as follows:

$$CO = HR \times SV$$

$$\frac{\text{Volume}}{\text{min}} = \frac{\text{beats}}{\text{min}} \times \frac{\text{volume}}{\text{beats}}$$

It should be evident from this relationship that all influences on cardiac output must act by changing either heart rate or stroke volume.

CONTROL OF HEART RATE

As discussed in Chap. 2, normal rhythmic contractions of the heart occur because of spontaneous electrical pacemaker activity of cells in the sinoatrial (SA) node. The interval between heartbeats (and thus the heart rate) is determined by how long it takes the membranes of these pacemaker cells to spontaneously depolarize to the threshold level. The heart beats at a spontaneous or *intrinsic rate* (≈ 100 beats per minute) in the absence of any outside influences. Outside influences *are* required, however, to increase or decrease the heart rate from its intrinsic level.

The two most important outside influences on heart rate come from the autonomic nervous system. Fibers from both the sympathetic and parasympathetic divisions of the autonomic system terminate on cells in the SA node and both can modify the intrinsic heart rate. Activating the cardiac sympathetic nerves (increasing cardiac sympathetic *tone*) increases the heart rate. Increasing cardiac parasympathetic tone slows the heart. As shown in Fig. 3-1, the parasympathetic and sympathetic nerves both influence heart rate by altering the course of spontaneous depolarization of the resting potential in SA pacemaker cells.

Cardiac parasympathetic fibers, which travel to the heart through the *vagus* nerves, release the transmitter substance *acetylcholine* on SA nodal cells. Acetylcholine increases the permeability of the resting membrane to K^+. As indicated in Fig. 3-1, this has two effects on the resting potential of cardiac pacemaker cells: (1) it causes an initial hyperpolarization of the resting membrane potential by bringing it closer to the K^+ equilibrium potential, and (2) it slows the rate of spontaneous depolarization of the resting membrane. Both these effects increase the time between beats by prolonging the time

Figure 3-1 Effect of sympathetic and parasympathetic tone on pacemaker potential.

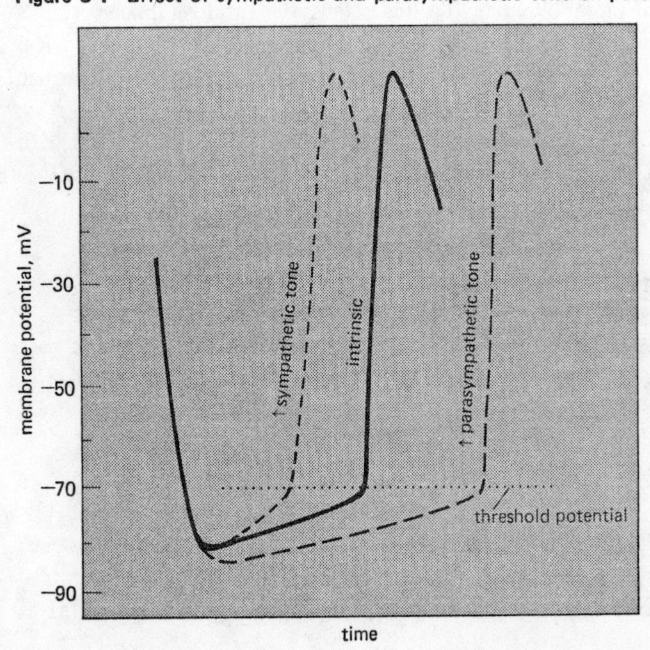

required for the resting membrane to depolarize to the threshold level. There is normally some continuous *tonic* activity of cardiac parasympathetic nerves, which causes the normal resting heart rate to be approximately 70 beats per minute.

Sympathetic nerves release the transmitter substance *norepinephrine* on cardiac cells. As shown in Fig. 3-1, norepinephrine increases heart rate by increasing the rate of depolarization of the resting membrane. The ionic bases for this are unclear at present but may involve changes in K^+, Na^+, and Ca^{2+} permeabilities as well as the sodium pump.

Besides their effect on heart rate, autonomic fibers also influence the conduction velocity of action potentials through the heart. Increases in sympathetic activity increase conduction velocity, whereas increases in parasympathetic activity decrease conduction velocity. These effects are most notable at the AV node and can influence the duration of the PR interval.

In addition to sympathetic and parasympathetic nerves, there are many, but usually less important, factors that can alter heart rate. These include a number of ions and circulating hormones, as well as physical influences such as temperature and atrial wall stretch. All act by somehow altering the time required for the resting membrane to depolarize to the threshold potential. An abnormally high concentration of Ca^{2+} in the extracellular fluid, for example, tends to decrease heart rate by shifting the threshold potential. Factors that increase heart rate are said to have a *positive chronotropic effect.* Those that decrease heart rate have a *negative chronotropic effect.*

CONTROL OF STROKE VOLUME

Starling's Law of the Heart

The volume of blood that the heart ejects with each beat can vary significantly. One of the most fundamental causes of variations in stroke volume was described by William Howell in 1884 and by Otto Frank in 1894, and was formally stated as the *law of the heart* by E. H. Starling in 1918. These investigators demonstrated that *the heart contracts more forcefully during systole when it is filled to a greater degree during diastole.* Figure 3-2, which summarizes Starling's findings, shows that increasing ventricular end-diastolic volume causes an increase in the pressure that the ventricle can develop during systole. This positive effect that ventricular filling has on the forcefulness of ventricular contraction is now referred to as Starling's law of the heart. To understand the basis for this observation, it is necessary to understand how the individual cardiac muscle cells are influenced by changes in the end-diastolic volume of the ventricle.

Law of Laplace Because of geometric factors, ventricular volume and pressure have a fixed relationship to the length and tension of cardiac muscle

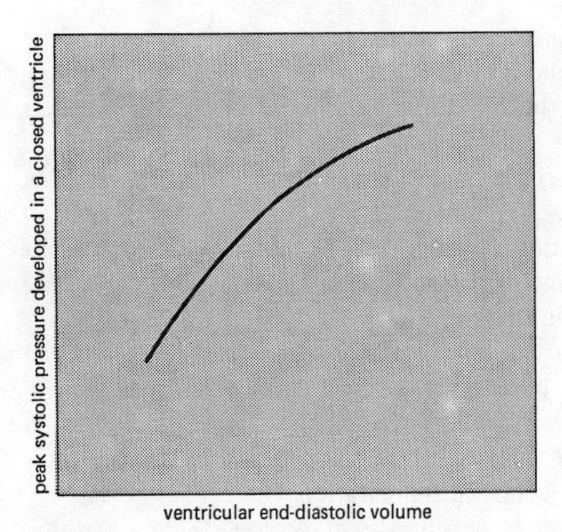

Figure 3-2 Starling's law of the heart.

fibers in the ventricular wall. The ventricle has a shape that, as a first approximation, is between a cylinder and a sphere. Each muscle fiber can be thought of as being oriented along the circumference of some circular path around the ventricular chamber, as illustrated in Fig. 3-3. Thus the length of an individual cardiac muscle fiber (L) is directly proportional to the radius (r) of the intraventricular chamber, since the circumference of a circle is directly proportional to its radius ($c = 2\pi r$). Moreover, ventricular volume has a fixed relationship to radius ($V \propto r^2$ for a cylinder and $V \propto r^3$ for a sphere). Thus each ventricular volume corresponds to a certain muscle cell length. Increasing ventricular end-diastolic volume, for example, increases end-diastolic muscle fiber length.

Increases in intraventricular pressure are *usually* accompanied by increases

Figure 3-3 Law of Laplace.

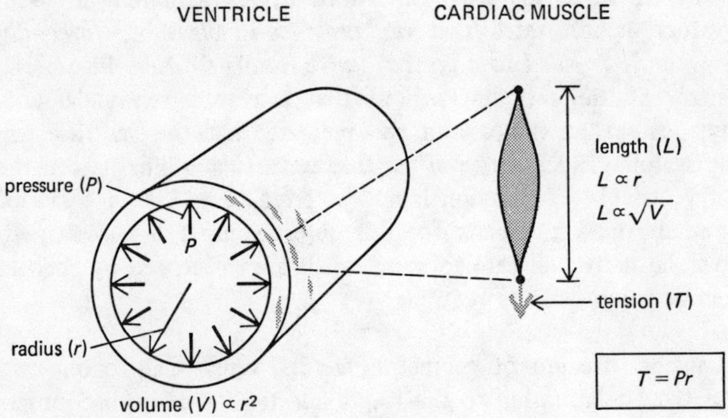

VENTRICLE

CARDIAC MUSCLE

length (L)

$L \propto r$

$L \propto \sqrt{V}$

pressure (P)

P

tension (T)

radius (r)

volume (V) $\propto r^2$

$T = Pr$

in muscle fiber tension and vice versa. The qualification "usually" must be used in the preceding sentence because, even when ventricular pressure is constant,. a change in ventricular size will change the tension on ventricular muscle fibers. This is because the total force that a pressure (force per unit area) exerts on a body is equal to the pressure times the surface area through which the pressure is acting. Thus the total force exerted on the ventricular walls by intraventricular pressure is equal to the intraventricular pressure times the surface area of the ventricular chamber. In addition, physical laws require that tensions within the ventricular walls be such that they balance the forces exerted on the walls by pressure. Wall tensions directed along the circumference of any circular path around the ventricle must be borne by muscle fibers oriented on that path. It can be shown, for any circular path on any vessel, that the *circumferential wall tension (T) must equal the internal pressure (P) times the internal radius (r); i.e.,* $T = Pr$. This relationship is called the *law of Laplace* and it has important cardiovascular implications with regard to both the heart and blood vessels.

Intraventricular pressure, ventricular radius, and ventricular wall tension must *always* be related by the law of Laplace. For example, we can see from the Laplace relationship how the tension on cardiac muscle fibers at the end of diastole is determined by end-diastolic volume and end-diastolic pressure. Moreover, the law of Laplace shows how, during systole, the tension developed by contracting muscle fibers generates pressure within the ventricle. Despite its continual and important role in governing ventricular function, the law of Laplace does not explain Starling's law of the heart. In fact, these two laws predict opposite effects of ventricular end-diastolic volume on ventricular pressure development. The law of Laplace says that as the ventricular radius gets larger, less pressure will be developed by a given increase in muscle tension ($P = T/r$). Howell, Frank, and Starling observed, however, that increasing ventricular end-diastolic volume increases the ventricle's ability to develop pressure and eject blood. The implication is that increasing ventricular size must increase the contractile ability of cardiac muscle fibers.

Pressure-Volume and Length-Tension Cycles What transpires after an action potential triggers a muscle contraction depends to a large extent on what is allowed to happen by the external constraints placed on the muscle. Activating a muscle whose ends are held rigidly causes it to develop tension, but it cannot shorten. This is called an isometric ("fixed-length") contraction. Activating an unrestrained muscle, on the other hand, causes it to shorten, but it will not develop tension because it has nothing to develop force against. This kind of contraction is called an isotonic ("fixed-tension") contraction. Thus, whether a muscle shortens or develops tension during contraction depends not on the muscle but on the external constraints placed on it.

The law of Laplace allows us to make some predictions about the length and tension changes that occur in the ventricular wall during a cardiac cycle. The relationship between intraventricular pressure and volume changes during a typical cardiac cycle is indicated in Fig. 3-4A, and the corresponding muscle length and tension changes are shown in Fig. 3-4B. (Figure 3-4A is simply another way of representing the intraventricular pressure and volume changes previously illustrated in Fig. 2-10. It is suggested that the student compare Fig. 3-4A with Fig. 2-10 until their interrelationship is clear.)

It is apparent from Fig. 3-4 that each major ventricular phase of the cardiac cycle has a corresponding phase of cardiac muscle length and tension change. During diastolic ventricular filling, for example, the progressive increases in ventricular pressure and volume combine to increase muscle tension ($T = Pr$), which passively stretches the resting cardiac muscle to greater

Figure 3-4 Ventricular pressure-volume cycle (A) and corresponding cardiac muscle length-tension cycle (B).

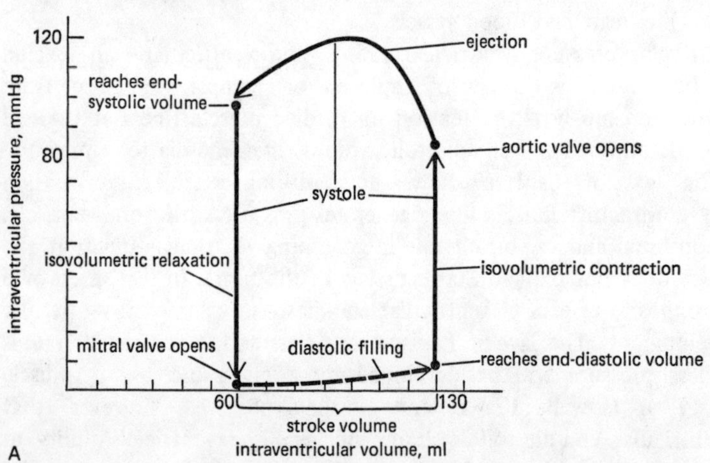

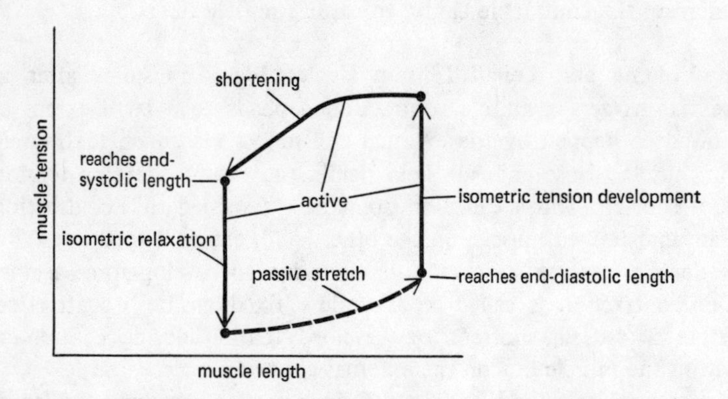

lengths. At the onset of systole, the ventricular muscle cells develop tension isometrically and intraventricular pressure rises accordingly ($P = T/r$). After the intraventricular pressure rises sufficiently to open the outlet valve, ventricular ejection begins as a consequence of ventricular muscle shortening. Thus isometric tension development normally occurs only for a brief period at the beginning of cardiac muscle contraction. During cardiac ejection, cardiac muscle is simultaneously generating active tension and shortening. The magnitude of ventricular volume change during ejection (or stroke volume) is determined simply by how far ventricular muscle cells shorten during contraction. Therefore, to understand how stroke volume is controlled, it is necessary to understand the factors that control the contractile ability of cardiac muscle itself.

Isometric Contractions: Length-Tension Relationships The relationships between muscle length and tension where the muscle is held at a fixed length are illustrated by the experimental procedure and observations shown in Fig. 3-5. The muscle is arranged so that its force can be measured at rest and during contraction at different lengths.

The first important fact illustrated in Fig. 3-5 is that force is required to stretch a resting muscle to different lengths. This force is called the *resting tension*. The lower curve in the graph in Fig. 3-5 shows the resting tension measured at different muscle lengths and is referred to as the *resting length-tension curve.* When a muscle is stimulated to contract while its length is held constant, it develops an additional component of tension called *active tension.* The *total tension* exerted by a muscle during contraction is the sum of the active and resting tensions.

The second important fact illustrated in Fig. 3-5 is that the active tension developed by cardiac muscle during the course of an isometric contraction depends very much on the muscle length at which the contraction occurs. Active tension development is maximum at some intermediate length L_{max}. Little active tension is developed at very short or very long muscle lengths. Normally, cardiac muscle operates at lengths well below L_{max}, so that increasing muscle length increases the tension developed during an isometric contraction.

It is believed that a muscle's ability to develop active tension inherently depends on muscle length because muscle length determines the extent of overlap of the thick and thin filaments in the sarcomere, as shown at the top of Fig. 3-5. For example, in the extreme case of very long muscle length, no active tension production is possible because there is no region of thick and thin filament overlap where cross-bridge formation can occur. At L_{max}, thin filaments from either end of the sarcomere overlap the thick filaments so that maximal cross-bridge formation is possible and the potential to develop active tension is maximal. At muscle lengths shorter than L_{max}, thin filaments from

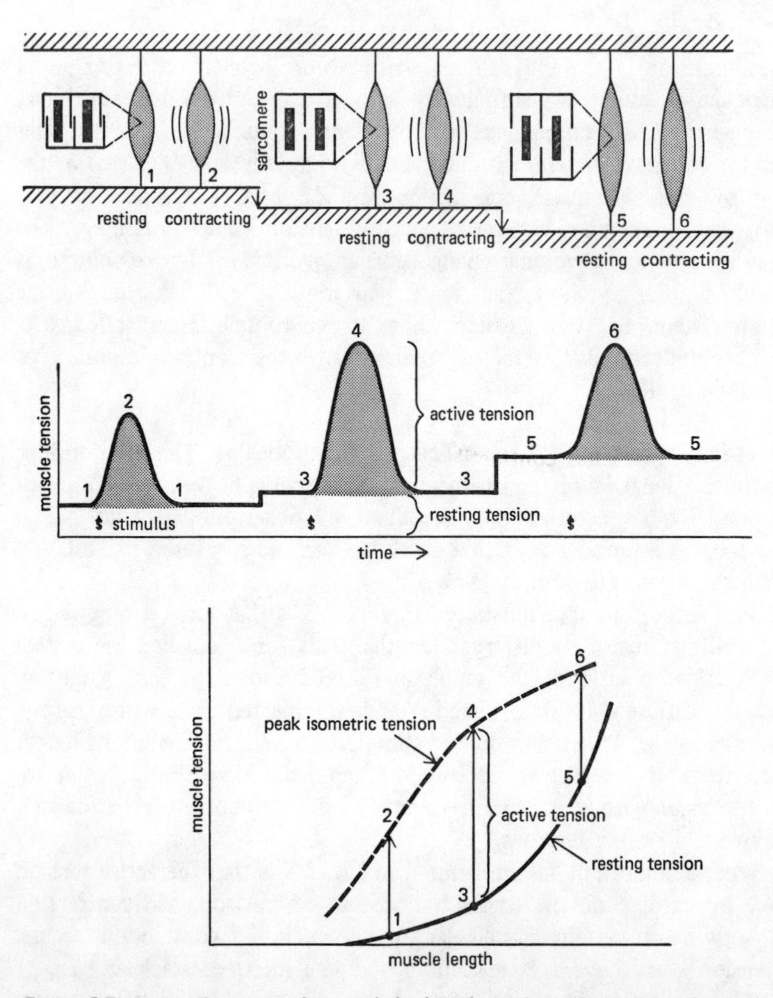

Figure 3-5 Isometric contractions and the length-tension relationship.

opposite ends of the sarcomere extend into the central region of the thick filaments. Presumably, cross-bridge formation is hampered or impossible in the central section of the thick filaments. Furthermore, thick filaments may abut on the Z line at very short muscle lengths.

The dependence of active tension development on muscle length is a fundamental property that has extremely important effects on heart function and forms the basis of Starling's law of the heart. Although the isometric length-tension relationship conveys important information about the effect of end-diastolic volume on the ventricle's ability to develop pressure, it is not immediately clear how stroke volume is affected by the pressure-generating ability of the ventricle, because cardiac muscle shortening is required for cardiac ejection. To fully appreciate how end-diastolic volume affects stroke volume and thus cardiac output, we must consider how cardiac muscle behaves when it is both developing tension *and* shortening.

Isotonic and Afterloaded Contractions During what is termed isotonic contraction, a muscle shortens against a constant load. A muscle contracts isotonically when lifting a fixed weight such as the 1-g load shown in Fig. 3-6. Recall that a 1-g tension placed on a resting muscle will result in some specific resting muscle length, which is determined by the muscle's resting length-tension curve. Recall also that if the muscle were to contract isometrically at this length, it would be capable of generating a certain amount of tension, e.g., 4.5 g as indicated by the dashed line in the graph of Fig. 3-6. A contractile potential of 4.5 g obviously cannot be realized by lifting a 1-g weight. When a muscle has contractile potential in excess of the tension it is actually developing, it shortens. Thus in an isotonic contraction muscle length decreases at constant tension, as illustrated by the horizontal arrow from

Figure 3-6 Relationship of isotonic and afterloaded contractions to the cardiac muscle length-tension diagram.

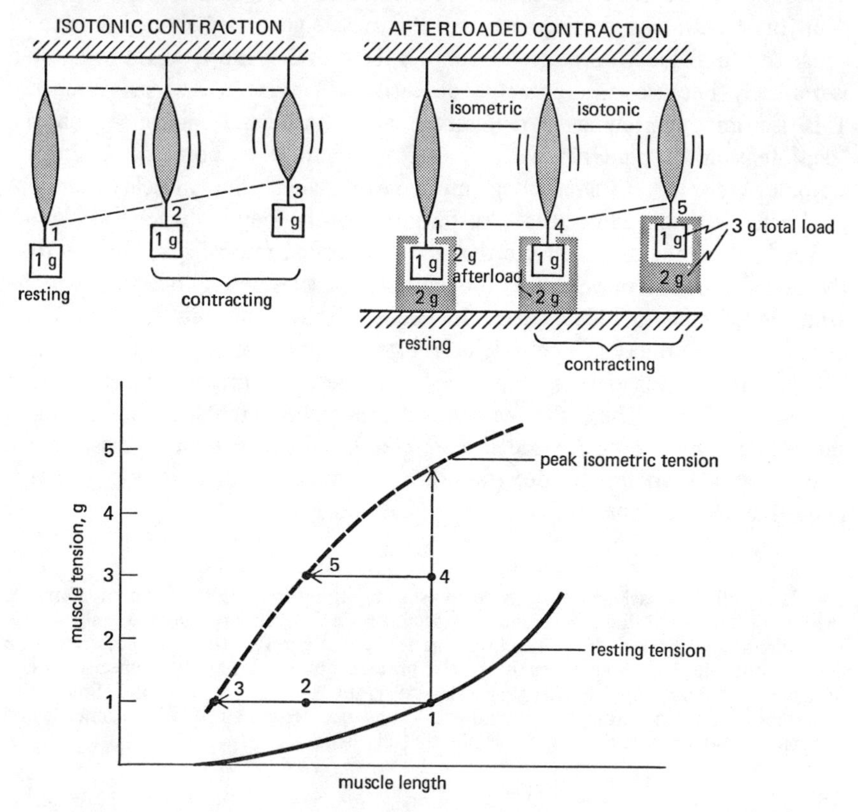

point 1 to point 3 in Fig. 3-6. As the muscle shortens, however, its contractile potential is inherently decreased, as indicated by the peak isometric tension curve in Fig. 3-6. There exists some short length at which the muscle is capable of generating only 1 g of tension, and when this length is reached shortening must cease.[1] Thus the curve on the cardiac muscle length-tension diagram that indicates how much isometric tension a muscle can develop at various lengths also establishes the limit on how far muscle shortening can proceed with different loads.

Figure 3-6 also shows a complex type of muscle contraction called an *afterloaded isotonic contraction*, in which the load on the muscle at rest, the *preload*, and the load on the muscle during contraction, the *total load*, are different. In the example of Fig. 3-6 the preload is equal to 1 g, and because an additional 2-g weight is engaged during contraction (the afterload), the total load equals 3 g.

Since preload determines the resting muscle length, both isotonic contractions shown in Fig. 3-6 begin from the same length. Because of the different loading arrangement, however, the afterloaded muscle must increase its total tension to 3 g before it can shorten. This initial tension will be developed isometrically and can be represented as going from point 1 to point 4 on the length-tension diagram. Once the muscle generates enough tension to equal the total load, its tension output is fixed at 3 g and it will now shorten isotonically because its contractile potential still exceeds its tension output. This isotonic shortening is represented as a horizontal movement on the length-tension diagram along the line from point 4 to point 5. As in any isotonic contraction, shortening must cease when the muscle's tension-producing potential is decreased sufficiently by the length change to be equal to the load on the muscle. Note that the afterloaded muscle shortens less than the nonafterloaded muscle even though both muscles began contracting at the same initial length. Note also, by comparison with Fig. 3-4, that an afterloaded contraction approximates closely how cardiac muscle actually behaves in the ventricle during the isovolumetric contraction and ejection phases of systole. Thus, the factors that affect the extent of cardiac muscle shortening during an afterloaded contraction are of special interest to us, because stroke volume is determined by how far cardiac muscle shortens under these conditions.

[1] In reality, muscle shortening requires some time and the duration of a muscle twitch contraction is limited because intracellular Ca^{2+} levels are elevated only briefly following the initiation of a membrane action potential. For this and possibly other reasons, isotonic shortening may not actually proceed quite as far as the isometric tension development curve on the length-tension diagram suggests is possible. Since this complication does not alter the general correspondence between a muscle's isometric and isotonic performance, it is neglected in the text discussion.

Ventricular Preload, Afterload, and Shortening How preload and total load affect the extent of cardiac muscle shortening can be appreciated from the length-tension diagrams of Fig. 3-7. Figure 3-7A illustrates how increasing muscle preload will increase the extent of shortening during a subsequent contraction with a fixed total load. Recall from the nature of the resting length-tension relationship that an increased preload is necessarily accompanied by increased initial muscle fiber length. When a muscle starts from a greater length, it has more room to shorten before it reaches the length at which its tension-generating capability equals the total load.[2] The pressure that fills the ventricle is often referred to as the *ventricular preload* because it controls end-diastolic volume, the preload on ventricular muscle fibers, and

[2] The initial speed of shortening against a fixed total load is also increased when initial muscle length is increased. This happens because increasing initial muscle length increases the tension-generating potential. This increases the initial shortening rate because it increases the amount by which the tension-generating potential exceeds the total load. A higher initial shortening rate makes it possible for more shortening to occur in the limited time in which the muscle remains activated.

Figure 3-7 Effect of changes in preload (A) and afterload (B) on cardiac muscle shortening during afterloaded contractions.

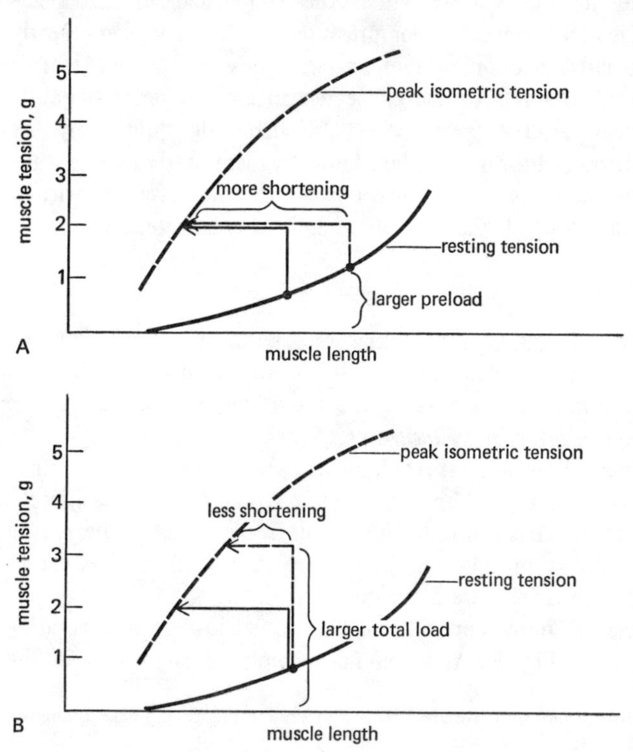

their initial length. Increases in ventricular filling pressure have a profound positive effect on stroke volume because longer initial fiber lengths greatly enhance the amount of muscle fiber shortening in an afterloaded contraction. It should be noted in Fig. 3-7A that increasing preload increases initial muscle length without significantly changing the final length to which the muscle shortens against a constant total load. Thus increasing ventricular filling pressure increases stroke volume primarily by increasing end-diastolic volume. This is not accompanied by a significant increase in end-systolic volume because the enhanced strength of contraction that comes from larger end-diastolic volume through Starling's law ensures that the extra blood that enters the ventricle during diastole is ejected during systole.

Figure 3-7B shows how increased total load, at constant preload, has a negative effect on cardiac muscle shortening. Again, this is simply a consequence of the fact that muscle cannot shorten beyond the length at which its isometric tension-generating potential equals the total load: the larger the total load, the greater the muscle length at which tension-generating potential equals it. Thus shortening must stop at a greater muscle length when total load is increased.

Recall from Fig. 3-4 that the intraventricular pressure during cardiac ejection reflects the tension against which the ventricular muscle fibers shorten. Since arterial and intraventricular pressures are normally very nearly identical during ventricular ejection, systemic arterial pressure is often referred to as left *ventricular afterload.*[3] Normally, mean ventricular afterload is quite constant, because mean arterial pressure is held within tight limits by the cardiovascular control mechanisms described later. In many pathological situations such as hypertension and aortic valve obstruction, however, ventricular function is adversely influenced by abnormally high ventricular afterload.

Cardiac Muscle Contractility

A number of factors in addition to initial muscle length can affect the tension-generating potential of cardiac muscle. *Any agent that increases the peak isometric tension that a muscle can develop at a* fixed length *is said to increase cardiac muscle contractility (a positive inotropic effect).*

The most important physiological regulator of cardiac muscle contractility is norepinephrine. When norepinephrine is released on cardiac muscle cells from sympathetic nerves, it has not only the chronotropic effect on heart rate discussed earlier but also a pronounced positive inotropic effect that causes cardiac muscle cells to contract more forcefully.

The positive effect of norepinephrine on the isometric tension-generating potential is illustrated in Fig. 3-8A. When norepinephrine is present in the

[3] This practice is somewhat unfortunate because arterial pressure is more analogous to ventricular total load than to ventricular afterload.

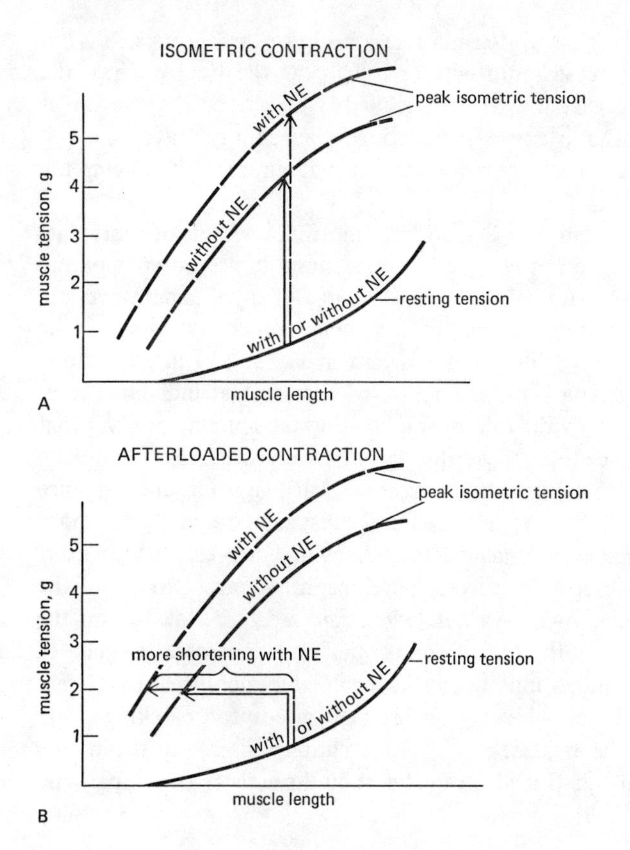

Figure 3-8 Effect of norepinephrine (NE) on isometric (A) and afterloaded (B) contractions of cardiac muscle.

solution bathing cardiac muscle, the muscle will, at every length, develop more isometric tension when stimulated than it would in the absence of norepinephrine. In short, norepinephrine raises the peak isometric tension curve on the cardiac muscle length-tension graph. Norepinephrine is said to increase cardiac muscle contractility because it enhances the forcefulness of muscle contraction even when length is constant. Changes in contractility and initial length can occur simultaneously, but by definition a change in contractility must involve a shift from one peak isometric length-tension curve to another.

Figure 3-8B shows how the ability of norepinephrine to raise the peak osmetric length-tension curve increases the amount of shortening in afterloaded contractions of cardiac muscle. With preload and total load constant, more shortening occurs in the presence of norepinephrine than in its absence. This is because when contractility is increased, the tension-generating potential is equal to the total load at a shorter muscle length. Note that norepinephrine has no effect on the resting length-tension relationship of cardiac muscle. Thus norepinephrine causes increased shortening by changing the final but not the initial muscle length associated with afterloaded contractions. In the intact heart, therefore, sympathetic nerve stimulation changes stroke volume primarily by decreasing the end-systolic volume without directly influencing the end-diastolic volume.

The intracellular mechanisms behind the inotropic effect of norepinephrine on cardiac muscle are not completely understood. Since norepinephrine can change the contractile strength at a fixed degree of myofilament overlap, it is presumed that it must exert its inotropic action by altering the percentage of available cross-bridge sites that are in use at any instant during contraction and/or the strength of interaction of filaments at individual sites. Norepinephrine is thought to do this by modulating the amount of Ca^{2+} that is released into the intracellular space during excitation-contraction coupling. Precisely how this is accomplished is unknown, although current evidence indicates the involvement of an intracellular substance called cyclic AMP (cyclic adenosine 3′,5′-monophosphate). It is known, however, that to exert both inotropic and chronotropic effects, norepinephrine must first combine with a specific receptor, known as a *beta-adrenergic receptor*, located on the cardiac cell membrane. Both the chronotropic and inotropic effects of sympathetic nerves on cardiac muscle can be completely abolished by certain specific chemicals called beta-blocking agents.[4] Beta-receptor blocking drugs are commonly used in the treatment of coronary artery disease to thwart the increased metabolic demands placed on the heart by the activity of sympathetic nerves.

CONTROL OF CARDIAC OUTPUT

The major influences on cardiac output that have been discussed in this chapter are summarized in Fig. 3-9. Heart rate is controlled by chronotropic influences on the spontaneous electrical activity of SA nodal cells. Cardiac parasympathetic nerves have a negative chronotropic effect and sympathetic nerves have a positive chronotropic effect on the SA node. Stroke volume is controlled by influences on the contractile performance of ventricular cardiac muscle—in particular its degree of shortening in the afterloaded situation. The three distinct influences on stroke volume are contractility, preload, and afterload. Increased cardiac sympathetic nerve activity tends to increase stroke volume by increasing the contractility of cardiac muscle. Increased arterial pressure tends to decrease stroke volume by increasing the afterload on cardiac muscle fibers. Increased ventricular filling pressure increases end-

[4] Circulating catecholamines also have positive chronotropic and inotropic effects on the heart that can be blocked with beta blockers. However, normal blood levels of catecholamines are so low that their effects on the heart are usually negligible.

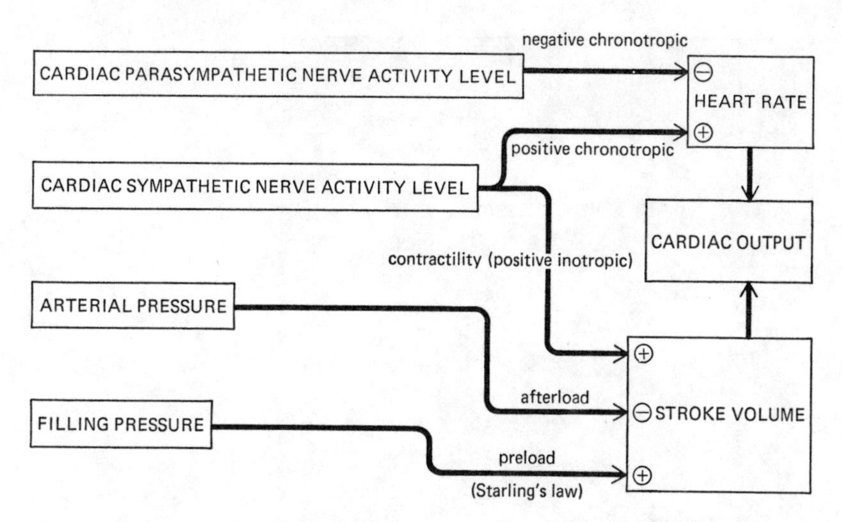

Figure 3-9 Influences on cardiac output.

diastolic volume, which tends to increase stroke volume through Starling's law.

It is important to recognize at this point that both heart rate and stroke volume are subject to more than one influence. Thus the fact that increased contractility tends to increase stroke volume should not be taken to mean that, in the intact cardiovascular system, stroke volume is always high when contractility is high. Following blood loss due to hemorrhage, for example, stroke volume may be low in spite of a high level of sympathetic nerve activity. Still, the information presented in this chapter is directly applicable to all cardiovascular situations including hemorrhage. We can correctly reason, for example, that a high level of cardiac sympathetic nerve activity cannot be the cause of the low stroke volume accompanying hemorrhage. The only possible causes for low stroke volume are high arterial pressure or low cardiac filling pressure. Since arterial pressure is normal or low following hemorrhage, the low stroke volume associated with severe blood loss must be (and is) the result of low cardiac filling pressure.

Cardiac Function Curves

One very useful way to summarize the influences on cardiac function and the interactions between them is by *cardiac function curves* such as those shown in Fig. 3-10. Cardiac output is treated as the dependent variable and is plotted on the vertical axis in Fig. 3-10. Cardiac filling pressure is plotted on the horizontal axis, and different curves are used to show the influence of changes in cardiac contractility associated with alterations in sympathetic nerve activity. Thus, Fig. 3-10 shows how the cardiac filling pressure and the

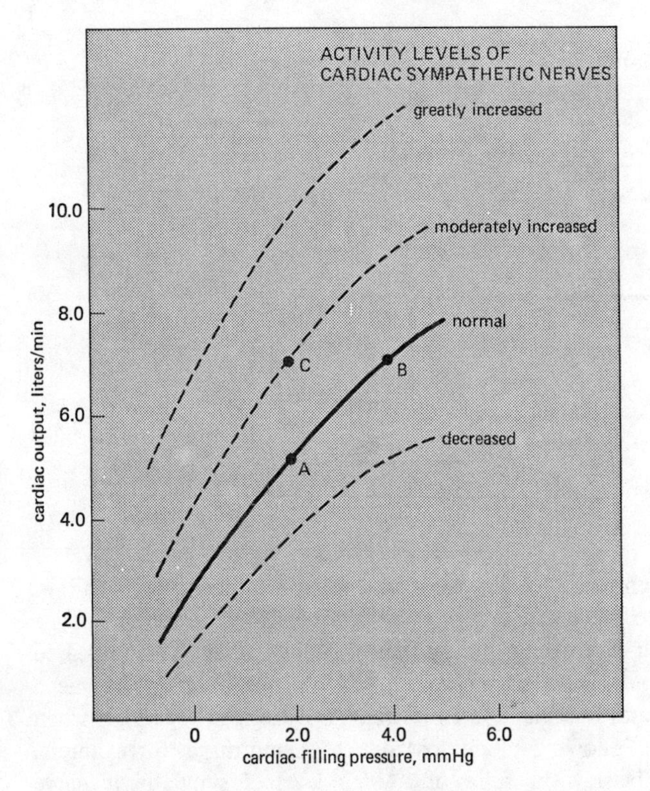

Figure 3-10 Influence of cardiac sympathetic nerves on cardiac function curves.

activity level of cardiac sympathetic nerves interact to determine cardiac output. When cardiac filling pressure is 2 mmHg and the activity of cardiac sympathetic nerves is normal, the heart will operate at point A and will have a cardiac output of 5 liters/min. Each single curve in Fig. 3-10 shows how cardiac output would be changed by changes in cardiac filling pressure if cardiac sympathetic nerve activity were held at a fixed level. For example, if cardiac sympathetic nerve activity remained normal, increasing cardiac filling pressure from 2 to 4 mmHg would cause the heart to shift its operation from point A to point B on the cardiac function diagram. In this case, cardiac output would increase from 5 to 7 liters/min solely as a result of the increased filling pressure (Starling's law). If, on the other hand, cardiac filling pressure were fixed at 2 mmHg while the activity of cardiac sympathetic nerves was moderately increased from normal, the heart would change from operating at point A to operating at point C. Cardiac output would again increase from 5 to 7 liters/min. In this instance, however, cardiac output does not increase through the length-dependent mechanism because cardiac filling pressure did not change. Cardiac output increases at constant filling pressure with an increase in cardiac sympathetic activity for two reasons. First, increased cardiac sympathetic nerve activity increases heart rate. Second, but just as important, increased sympathetic nerve activity increases stroke volume by increasing cardiac contractility. Cardiac function graphs thus consolidate our knowledge of many mechanisms of cardiac control, and we will find them most helpful in understanding how the heart interacts with other elements in the cardiovascular system.

MEASUREMENT OF CARDIAC OUTPUT

There are a number of clinical methods of measuring cardiac output that use the Fick principle discussed in Chap. 1. For calculating blood flow, the Fick equation can be rearranged as follows:

$$\dot{Q} = \frac{\dot{X}_{\text{tc}}}{[\text{X}]_a - [\text{X}]_v}$$

A common method of determining cardiac output is to use the Fick principle to calculate the collective flow through the systemic organs from (1) the whole body oxygen consumption rate ($\dot{X}_{\text{tc}}$), (2) the oxygen concentration in arterial blood ($[\text{X}]_a$), and (3) the concentration of oxygen in mixed venous blood ($[\text{X}]_v$). Of the values required for this calculation, the oxygen content of mixed venous blood is the most difficult to obtain. Generally, the sample for venous blood oxygen measurement must be taken from venous catheters positioned in the right ventricle or pulmonary artery to ensure that it is a mixed sample of venous blood from all systemic organs.

The calculation of cardiac output from the Fick principle is best illustrated by an example. Suppose a patient is consuming 250 ml of O_2 per minute when his or her systemic arterial blood contains 200 ml of O_2 per liter and the right ventricular blood contains 150 ml of O_2 per liter. This means that, on the average, each liter of blood loses 50 ml of O_2 as it passes through the systemic organs. In order for 250 ml of O_2 to be consumed per minute, 5 liters of blood must pass through the systemic circulation each minute:

$$\dot{Q} = \frac{250 \text{ ml } O_2/\text{min}}{(200 - 150) \text{ ml } O_2/\text{liter blood}}$$

$$\dot{Q} = 5 \text{ liters blood/min}$$

Dye dilution and thermal dilution (dilution of heat) are other clinical techniques commonly employed for estimating cardiac output. Usually a known quantity of indicator (dye or heat) is rapidly injected into the blood as it enters the right heart and appropriate detectors are arranged to continuously record the concentration of the indicator in blood as it leaves the left heart. It is possible to estimate the cardiac output from the quantity of indicator injected and the time record of indicator concentration in the blood that leaves the left heart.

Study questions: **15** to **18**

Chapter 4

Components of the Peripheral Vascular System

OBJECTIVES

The student understands the physical factors that regulate blood flow through the various components of the vasculature:

1. Lists the major different types of vessels in a vascular bed and describes the morphological differences among them.
2. Describes differences in the blood flow velocity in the various segments and how these differences are related to their total cross-sectional area.
3. Describes laminar and turbulent flow patterns and the origin of flow sounds in the cardiovascular system.
4. Describes the pressure changes that occur as blood flows through a vascular bed and relates them to the vascular resistance of the various vascular segments.
5. States how the resistance of each consecutive vascular segment contributes to an organ's overall vascular resistance and, given data, calculates the overall resistance.

6 Defines total peripheral resistance and states the relationship between it and the vascular resistance of each systemic organ.
7 Identifies the approximate percentage of the total blood volume that is contained in the various vascular segments in the systemic circulation.
8 Defines vascular compliance and states how the volume-pressure curves for arteries and veins differ.
9 Predicts what will happen to venous volume when venous smooth muscle is activated or venous pressure is changed.
10 Defines peripheral venous pool and central venous pool.
11 Describes the role of arterial compliance in storing energy for blood circulation.

This chapter will describe the overall structural design of the vascular system and discuss the functional implications of this design. Much of this discussion also applies to the pulmonary vascular bed; the main exception is that the pulmonary arterial pressure is much lower than the systemic arterial pressure.

VASCULAR ARCHITECTURE

Blood that is ejected into the aorta by the left heart passes consecutively through many different types of vessels before it returns to the right heart. As diagrammed in Fig. 4-1, the major vessel classifications are *arteries, arterioles, capillaries, venules,* and *veins.* These consecutive vascular segments are distinguished from one another by differences in physical dimensions, morphological characteristics, and function. Some representative physical characteristics are shown in Fig. 4-1 for each of the major vessel types. It should be realized, however, that vascular beds are a continuum and that the transition from one type of vascular segment to another does not occur abruptly. The number and collective cross-sectional area values in Fig. 4-1 are estimates for the entire systemic circulation.

Arteries are thick walled vessels that contain, in addition to smooth muscle, a large amount of collagen and elastin fibers. These elastic elements make arteries relatively resistant to passive expansion when the pressure inside them increases. The aorta is the largest artery and has an inside diameter of about 25 mm. Arterial diameter decreases with each consecutive branching, and the smallest arteries have diameters of approximately 0.1 mm. The consecutive arterial branching pattern causes a geometric increase in arterial numbers to accompany the progressive decrease in size of individual arteries. The collective cross-sectional area increases from about 4.5 cm^2 at the root of the aorta to a total area of more than 100 cm^2 at the level of the smallest arteries. The numbers given in Fig. 4-1 represent approximate average values for each vessel type.

Arterioles are smaller and structured differently from arteries. In propor-

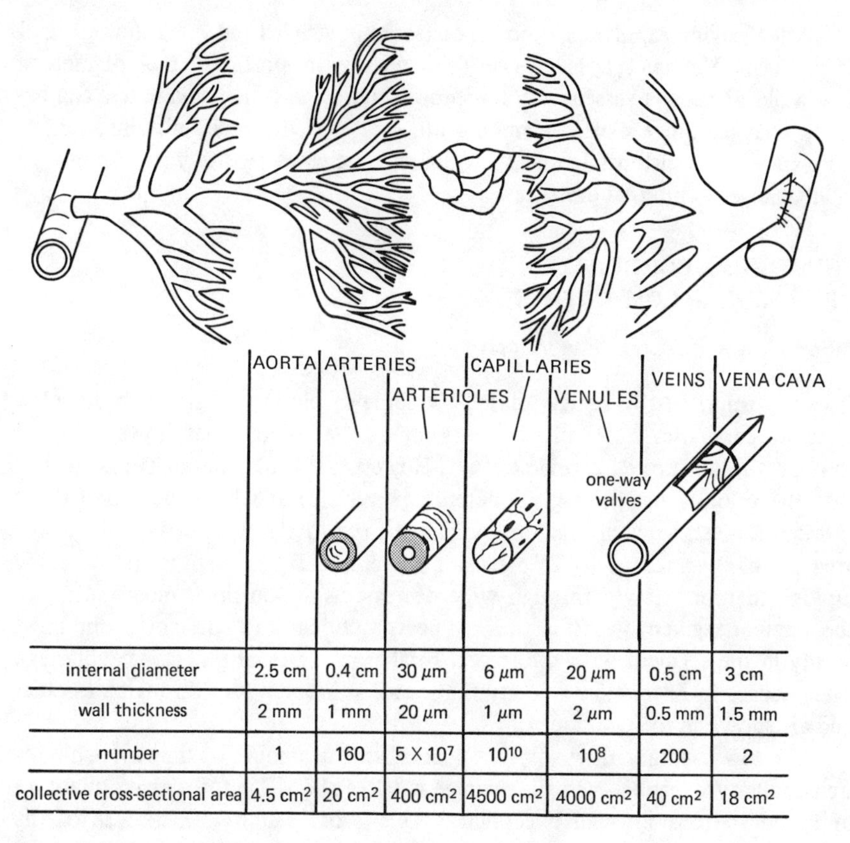

	AORTA	ARTERIES	ARTERIOLES	CAPILLARIES	VENULES	VEINS	VENA CAVA
internal diameter	2.5 cm	0.4 cm	30 μm	6 μm	20 μm	0.5 cm	3 cm
wall thickness	2 mm	1 mm	20 μm	1 μm	2 μm	0.5 mm	1.5 mm
number	1	160	5×10^7	10^{10}	10^8	200	2
collective cross-sectional area	4.5 cm^2	20 cm^2	400 cm^2	4500 cm^2	4000 cm^2	40 cm^2	18 cm^2

Figure 4-1 Structural characteristics of the peripheral vascular system.

tion to lumen size, arterioles have much thicker walls with more smooth muscle and less elastic material than arteries. Despite their minute size (a single sheet of paper is about 150 μm thick), arterioles are so numerous that in parallel their collective cross-sectional area is much larger than that at any level in arteries.

Capillaries are the smallest vessels in the vasculature. In fact, red blood cells with diameters of 7 μm must deform to pass through them. The capillary wall consists of a single layer of endothelial cells, which separate the blood from the interstitial fluid by only about 1 μm. Capillaries contain no smooth muscle and thus lack the ability to change their diameter actively. They are so numerous that the total collective cross-sectional area of all the capillaries in systemic organs is more than 1000 times that of the root of the aorta. Given that capillaries are about 0.5 mm in length, we can calculate that the total surface area available for exchange of material between blood and interstitial fluid is approximately 100 m^2.

After leaving capillaries, blood is collected in venules and veins and returned to the heart. Venous vessels have very thin walls in proportion to their diameters. The walls of venous vessels contain smooth muscle and their diameters can actively change. Since elastic elements are sparse in the walls of veins, venous vessels are quite distensible and their diameters change passively in response to small changes in internal pressure.

FUNCTIONAL CORRELATES OF VASCULAR COMPONENTS

Blood Flow and Blood Flow Velocity

It is very important to realize that the volume of blood that flows through the systemic capillaries in 1 min is the same as the volume that flows into and through the aorta, i.e., cardiac output. However, it is also important to realize that the velocity or speed (millimeters per minute) at which the blood flows through a given vascular segment varies with the total collective cross-sectional area of that segment. Just as a stream moves rapidly through narrow shallow rapids and more slowly through wide deep pools, blood flows most rapidly in the vascular segment with the smallest cross-sectional area (the aorta) and most slowly in the segment with the largest total cross-sectional area (the capillaries). The changes in linear flow velocity that occur as blood passes through a vascular bed are shown in the top trace of Fig. 4-2.

The low capillary flow velocity maximizes the amount of time available for transcapillary exchange. On the average it takes about 1 s for a given quantum of blood to pass through a capillary. This is only about one-sixtieth of the average time it takes the blood to make a complete circuit of the cardiovascular system.

Blood normally flows through all vessels in the cardiovascular system in an orderly streamlined manner called *laminar flow.* With laminar flow blood moves more rapidly in the center of the tube than near the vessel wall. Concentric layers of fluid move smoothly beside one another, and there is little mixing between fluid layers. When, however, blood is forced to move with high velocity through a narrow opening, the normal laminar flow pattern may break down into a *turbulent flow* pattern. With turbulent flow there is much mixing and friction between fluid layers. Turbulent flow also generates sound, which can be heard with the aid of a stethoscope. For example, cardiac murmurs are manifestations of turbulent flow patterns generated by cardiac valve abnormalities. Detection of sounds from peripheral arteries is abnormal and usually indicates significant narrowing of the lumen.

Blood Pressure

Blood pressure decreases in the consecutive segments, as shown in the second trace of Fig. 4-2. Recall from Fig. 2-10 that aortic pressure fluctuates between

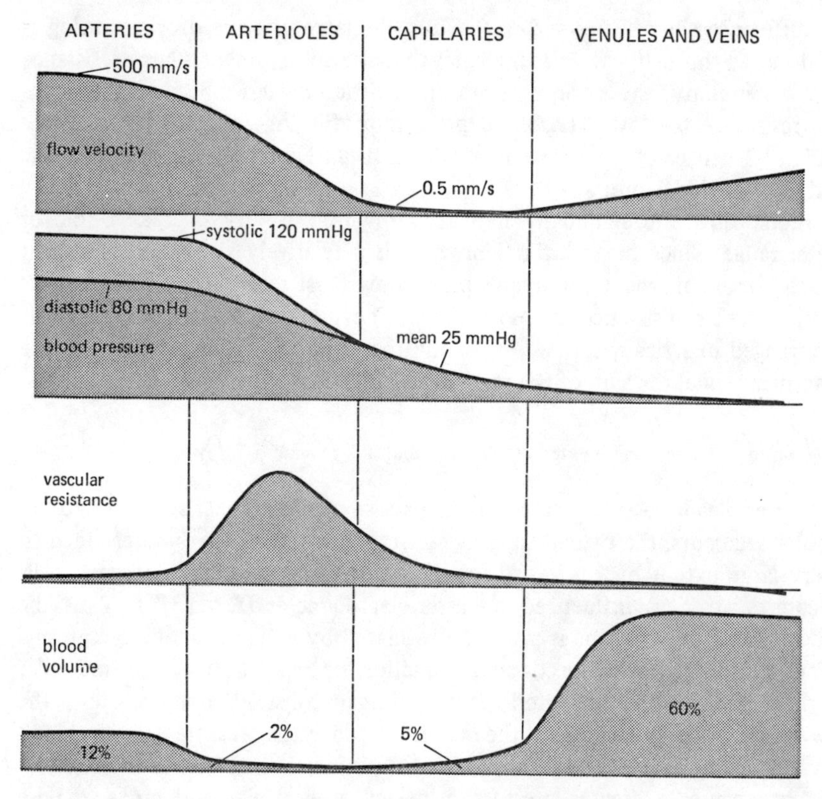

Figure 4-2 Functional correlates of peripheral vascular structure.

a systolic and diastolic value with each heartbeat. The average aortic pressure is normally about 100 mmHg and this *mean arterial pressure* drops very little as the blood passes through the arteries.

A large pressure drop occurs in the arterioles, where in addition the pulsatile nature of the pressure nearly disappears. Pressure continues to fall as blood moves through the capillaries. Whereas blood pressure is higher at the arterial end than at the venous end, mean capillary pressure is approximately 25 mmHg. Pressure continues to decrease in the venules and veins as blood returns to the right heart. The central venous pressure (which is the filling pressure for the right heart) is normally very close to 0 mmHg.

Vascular Resistance

Recall from Chap. 1 that resistance, pressure difference, and flow of fluid in a tube are related by the basic flow equation: $\dot{Q} = \Delta P/R$. Since the flow ($\dot{Q}$) must be the same through each consecutive segment of the vascular bed, the resistance of each consecutive vascular segment is directly related to the pressure difference across it ($R = \Delta P/\dot{Q}$). Thus, the large pressure drop occurring as blood moves through arterioles indicates that arterioles present a large resistance to flow. Similarly, the mean pressure drops little in arteries because they have little resistance to flow. The blood pressure profile shown in the second trace of Fig. 4-2 can exist only if the vascular resistance profile is as shown in the third trace of this figure.

Blood flow through many individual organs can vary over a 10-fold or greater range. Since mean arterial pressure is a relatively stable cardiovascular variable, large changes in an organ's blood flow must result from changes in its overall vascular resistance to blood flow. The consecutive vascular segments are arranged in series within an organ and the overall vascular resistance of the organ must equal the sum of the resistances of its consecutive vascular segments:

$$R_{\text{organ}} = R_{\text{arteries}} + R_{\text{arterioles}} + R_{\text{capillaries}} + R_{\text{venules}} + R_{\text{veins}}$$

Since arterioles have such a large vascular resistance in comparison to the other vascular segments, the overall vascular resistance of any organ is determined to a very large extent by the resistance of its arterioles. Arteriolar resistance is, of course, strongly influenced by arteriolar diameter ($R \propto 1/d^4$). Thus the blood flow through organs is primarily regulated by adjustments in the internal diameter of arterioles caused by contraction or relaxation of the muscular arteriolar walls. When the arterioles of an organ constrict, not only does the flow to the organ decrease, but the manner in which the pressure drops as blood flows through the organ is also modified. In general, an increase in arteriolar resistance causes a greater pressure drop across arterioles and tends to raise the pressure in arteries while lowering the pressure in capillaries and veins.

A logical question at this point might be why the arteriolar segment of a vascular bed has such a large resistance to flow when the capillary bed, which has vessels with much smaller diameters, has so little. The answer lies in the fact that capillaries are much more numerous than arterioles. When a number of individual resistive elements (R_1, R_2, . . . , R_n) are arranged in *parallel*, the total resistance (R_{total}) of the network is always *less* than that of any individual element in the group. Specifically,

$$\frac{1}{R_{\text{total}}} = \frac{1}{R_1} + \frac{1}{R_2} + \cdots + \frac{1}{R_n}$$

It can be shown from this equation that increasing the number of elements of equal resistance placed in parallel lowers the total resistance of the network. Thus the total resistance to flow through a capillary bed can be quite low, although individual capillaries have high vascular resistance. Arterioles are also generally arranged in parallel, and consequently the resistance to blood flow across the arteriolar segment of a vascular bed is much less than that of any individual arteriole. Since capillaries are several hundred times more numerous than arterioles, several hundred times more terms must be used when applying the equation above to the capillary bed than when applying it to the arteriolar segment of a vascular bed. The result is that total capillary segment resistance is less than total arteriolar segment resistance.

The parallel resistance equation also has important implications for the overall operation of the cardiovascular system, since the systemic organs are generally arranged in parallel (Fig. 1-2). *Total peripheral resistance* is the cardiovascular variable that indicates the overall resistance to blood flow through the entire systemic vascular system. The vascular resistance of each systemic organ contributes to total peripheral resistance as indicated in the parallel resistance equation. Thus total peripheral resistance is always lower than that of any systemic organ.

It should be noted, however, that increasing the resistance of any single element in a resistive network, whether arranged in series or in parallel, increases the total resistance of the system. Increasing renal vascular resistance, for example, will increase total peripheral resistance even though the kidney lies in parallel with other systemic organs.

Blood Volume

The bottom trace in Fig. 4-2 shows the approximate percentage of the total circulating blood volume that is contained in the different vascular segments of the systemic organs at any instant of time. Approximately 20 percent of the total volume is contained in the pulmonary system and the heart chambers and is not accounted for in this figure.

Note that only about 2 percent of the blood is in arterioles at any instant. If, for example, the arterioles of skeletal muscle were to constrict their diameter to one-half normal, the volume of blood contained in the arterioles would decrease to one-half normal. This would not change the volume distribution within the muscle very much. However, this amount of arteriolar constriction would increase resistance to flow 16-fold and consequently produce a dramatic reduction in the muscle's blood flow.

Most of the circulating blood is contained at any instant in the peripheral veins of the various peripheral organs. This diffuse but large blood reservoir is often referred to as the *peripheral venous pool.* A second but smaller reservoir of venous blood, called the *central venous pool,* is contained in the great veins of the thorax and the right atrium. When peripheral veins constrict, blood is displaced from the peripheral venous pool and enters the central pool. An increase in the central venous volume, and thus pressure, enhances cardiac filling, which in turn augments stroke volume according to Starling's law of the heart. This is an extremely important mechanism in cardiovascular regulation and will be discussed in Chap. 6.

It should be noted that whereas venous constrictions may displace a significant amount of blood from the veins of an organ, a decrease in the diameter of the veins will have little direct effect on the organ's overall vascular resistance. It should also be kept in mind that the volume of blood contained *within* a vessel segment and the rate of blood flow *through* it are independent. Identical quantities of blood must pass through all the vessel segments of an organ in 1 min.

Elastic Properties of Vessels

The elastic properties of vessels are often represented by a variable called *compliance* (C), which describes how much the volume will change (ΔV) with an increment change in internal pressure (ΔP):

$$C = \frac{\Delta V}{\Delta P}$$

Veins are much more compliant than arteries and undergo significant volume changes in response to only slight changes in pressure, as shown in Fig. 4-3. Standing upright, for example, increases venous pressure in the lower extremities and promotes blood accumulation (pooling) in these vessels. This might be represented as a shift from point A to point B in Fig. 4-3.

The dashed line in Fig. 4-3 indicates how venous compliance is altered when

Figure 4-3 Vascular pressure-volume curves.

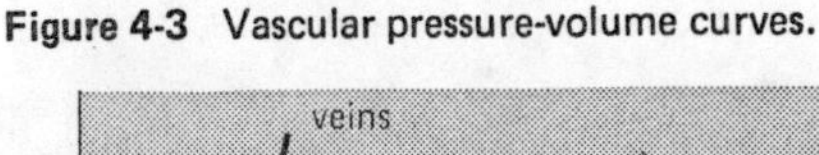

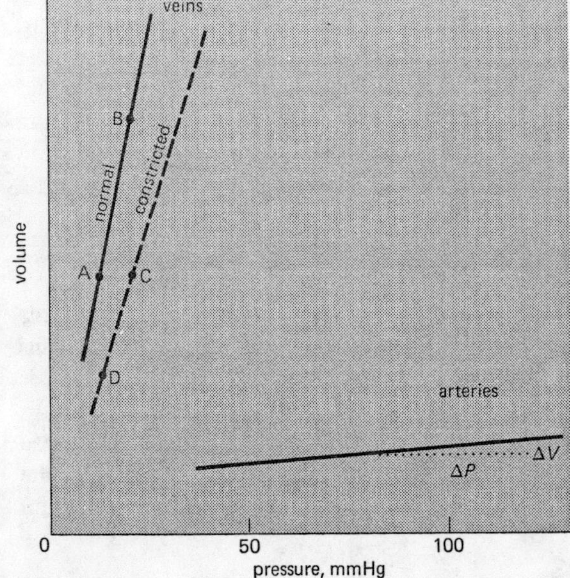

venous smooth muscle is activated and veins constrict. A shift from point B to point C illustrates how a venous volume increase due to increased venous pressure might be counteracted by activating venous smooth muscle and contracting veins. Similarly, a shift from point A to point D might indicate how the peripheral venous volume can be decreased by venous constriction when it is appropriate to shift blood into the central venous pool.

Arteries are much less compliant than veins and large pressure changes are required to produce small changes in the arterial volume. Recall that arterial pressure rises during each period of cardiac ejection and falls during diastole. The arterial pressure-volume curve of Fig. 4-3 indicates that fluctuations in arterial pressure must be accompanied by small but corresponding changes in arterial volume; arterial walls are stretched during systole and recoil during diastole. It is the recoil of the arterial walls that maintains the pressure needed to propel the blood through the peripheral vascular beds during diastole. Thus the elastic properties of the arterial walls play an important role in enabling blood to flow continuously through capillaries even through cardiac ejection is intermittent.

Study questions: **19** to **23**

Chapter 5

Peripheral Vascular Control

OBJECTIVES

The student understands the general mechanisms involved in local vascular control:

1. Defines vascular tone and describes at least two mechanisms by which tone may be produced.
2. Defines intrinsic tone.
3. States the local metabolic vasodilator hypothesis.
4. Lists several substances potentially involved in local metabolic control.
5. Defines active and reactive hyperemia and indicates a possible mechanism for each.
6. Defines autoregulation of blood flow and briefly describes the metabolic and myogenic theories of autoregulation.
7. Lists the major steps in the chain of events between an increase in sympathetic vasoconstrictor nerve fiber activity and vasoconstriction.
8. Distinguishes neurogenic tone from intrinsic tone.

9 Describes how vascular tone is influenced by epinephrine, acetylcholine, antidiuretic hormone, serotonin, histamine, bradykinin, and angiotensin II.
10 Lists the major influences on venous volume.
11 Describes in general how control of flow differs between organs with normally high intrinsic arteriolar tone and organs with normally high neurogenic arteriolar tone.

The student knows the dominant mechanisms of flow and blood volume control in the major body organs:

12 States the relative importance of local metabolic and nervous control of coronary blood flow.
13 Identifies the factors that determine the external work of the heart.
14 States how myocardial oxygen consumption is affected by arterial pressure, cardiac output, heart rate, cardiac contractility, ventricular wall tension, and cardiac dilation.
15 Defines systolic compression and indicates its relative importance to blood flow in the endocardial and epicardial regions of the right and left ventricular walls.
16 Describes the major mechanisms of flow and blood volume control in each of the following specific systemic organs: skeletal muscle, brain, splanchnic organs, skin, and kidney.
17 States why mean pulmonary arterial pressure is lower than mean systemic arterial pressure.
18 Describes how pulmonary vascular control differs from that in systemic organs.
19 Describes the factors that govern pulmonary transcapillary fluid movement and indicates their normal values.

VASCULAR SMOOTH MUSCLE

All blood vessels except capillaries are capable of active adjustments in their diameter because they have smooth muscle cells circumferentially oriented in their walls that can contract and relax. Although smooth muscle cells do not have the highly structured internal sarcomere arrangement that is typical of skeletal and cardiac muscles, they do contain thick and thin filaments and the contractile proteins actin and myosin. It is also clear that Ca^{2+} ions play an important role in regulating the interaction between the contractile proteins in smooth muscle and that changes in the intracellular Ca^{2+} concentration accompany changes in membrane potential. Thus, although smooth muscle contracts and relaxes much more slowly than either cardiac or skeletal muscle, the fundamental processes of contraction appear to be similar for all muscle types.

Vascular Tone

Vascular tone is a term used to indicate the general contractile state of a vessel or a vascular region. A vessel's level of vascular tone is determined by the level of activation of the smooth muscle cells. Increasing the level of activation of vascular smooth muscle cells increases the level of vascular tone. Vascular tone can be modulated and sustained at any level between those corresponding to complete inactivation and maximal activation of vascular smooth muscle.

Various mechanisms appear to be responsible for vascular tone in different vessels. Pacemaker activity and spontaneous action potential generation similar to those characteristic of cardiac muscle have been demonstrated for the vascular smooth muscle cells in some vessels. Thus vascular tone in these vessels may be the summation of the continual but independent rhythmic activations of the individual smooth muscle cells in the vessel wall. The inherently slow mechanical contraction and relaxation of individual muscle cells facilitates the blending of many desynchronized twitch contractions into a sustained vessel tone.

Some vessels, however, contract rhythmically rather than produce a sustained tone. The vascular smooth muscle in these vessels possesses not only pacemaker activity but also the ability to propagate electrical activity from cell to cell. Rhythmic fluctuations in vessel diameter and blood flow are commonly observed in microscopic studies of the circulation and have been termed *vasomotion.* Vasomotion is most prominent in the smallest arterioles near the capillary bed. Whereas it is natural to picture the steady flow through a whole organ as resulting from uniform flow through all its individual arterioles, this may be far from the case. What we think of and calculate as arteriolar resistance ($R = \Delta P/\dot{Q}$) may be correct only in a statistical sense for large numbers of arterioles and not indicative of the state of any particular arteriole.

Finally, in some vessels tone is generated in association with slow, low-amplitude depolarization of the "resting" membrane of smooth muscle cells. Action potentials are totally absent, and the level of contractile state corresponds roughly to the degree of depolarization of the resting membrane.

Although the mechanisms for vascular tone may vary from vessel to vessel, tone is ultimately related to the level of activation of vascular smooth muscle. Therefore, all *active* vascular responses are the result of changes in vascular tone.

CONTROL OF ARTERIOLAR TONE

As described in Chap. 4, the blood flow through any organ is determined largely by the vascular resistance, which is dependent on the diameter of the arterioles. Consequently, an organ's flow is controlled by factors that influence the arteriolar smooth muscle tone.

Arterioles remain in a state of partial constriction even when all external influences on them have been removed; hence they are said to have a degree of *intrinsic basal tone.* This intrinsic basal tone establishes a baseline of partial arteriolar constriction on which the external influences on arterioles exert their dilating or constricting effects. These influences can be separated into three categories: neural influences, local influences, and hormonal influences.

Neural Influences on Arterioles

Sympathetic vasoconstrictor fibers innervate arterioles in all systemic organs and provide the most important means of *reflex* control of the peripheral vasculature. These nerves release norepinephrine from their terminal structures in amounts proportional to their activity. Norepinephrine causes an increase in the tone of arterioles after combining with an *alpha-adrenergic receptor* on smooth muscle cells.

Sympathetic vasoconstrictor nerves normally have a continual or *tonic activity*. This tonic activity of sympathetic vasoconstrictor nerves makes the normal tone of arterioles considerably greater than their intrinsic tone. The additional component of vascular tone is called *neurogenic tone.* When the activity of sympathetic vasoconstrictor nerves is increased above normal, arterioles constrict and cause organ blood flow to fall below normal. On the other hand, vasodilation and increased organ blood flow can be caused by sympathetic vasoconstrictor nerves if their normal tonic activity level is reduced. Thus an organ's blood flow can either be reduced below normal or be increased above normal by changes in the sympathetic vasoconstrictor fiber tone.

As a general rule, blood vessels do not receive innervation from the parasympathetic division of the autonomic nervous system. The most prominent exceptions are the vessels of the brain and heart, where the presence of parasympathetic nerves is apparently inconsequential, and the external genitalia, where they release *acetylcholine*, which produces the vasodilation responsible for erection. Another exception to the general rules of vascular innervation occurs in the skeletal muscle in at least some species, where *sympathetic vasodilator* nerves that release acetylcholine have been demonstrated. It has not been established that similar nerves exist in humans. Even in species where they do exist the importance of cholinergic sympathetic vasodilator fibers is unclear, since exercise hyperemia proceeds quite normally after the vascular receptors for acetylcholine have been blocked. It is speculated that these fibers might play a role in increasing skeletal muscle blood flow in anticipation of exercise.

Local Mechanisms

The arterioles that control flow through a given organ lie within the organ tissue itself. Thus arterioles and the smooth muscle in their walls are exposed to the chemical composition of the interstitial fluid of the organ they serve. The interstitial concentrations of many substances reflect the balance between the metabolic activity of the tissue and its blood supply. Interstitial oxygen levels, for example, fall whenever the tissue cells are utilizing oxygen faster than it is being supplied to the tissue by blood flow. Conversely, interstitial oxygen levels rise whenever excess oxygen is being delivered to a tissue from the blood. In the peripheral vascular beds, exposure to low oxygen reduces arteriolar tone and causes an increase in arteriolar diameter (vasodilation), whereas high oxygen levels cause arteriolar vasoconstriction. Thus a local feedback mechanism exists that automatically operates on arterioles to regulate a tissue's blood flow in accordance with its metabolic needs. Whenever blood flow and oxygen delivery fall below a tissue's oxygen demands, the oxygen levels around arterioles fall, the arterioles dilate and reduce their resistance to blood flow, and the blood flow through the organ appropriately increases.

Many substances besides oxygen are found within tissues and can affect the tone of vascular smooth muscle. When the metabolic rate of skeletal muscle is increased by exercise, for example, not only do tissue levels of O_2 decrease, but those of CO_2, H^+, and K^+ increase. Muscle tissue osmolarity also increases during exercise. In addition, under circumstances of severe oxygen lack, muscle cells have been shown to release adenosine, which is an extremely potent vasodilator.

At present we do not know which of these (or possibly other) metabolically related chemical alterations within tissues are primarily responsible for local control of blood flow. It appears likely that arteriolar tone depends on the combined action of many factors. In addition, any given factor may have different degrees of importance in the local control of flow in different organs.

For conceptual purposes, our understanding of local metabolic control can be summarized as shown in Fig. 5-1. Vasodilator factors enter the interstitial space from the tissue cells at a rate proportional to tissue metabolism. These vasodilator factors are removed from the tissue at a rate proportional to blood flow. Whenever tissue metabolism is proceeding at a rate for which the blood flow is inadequate, the interstitial vasodilator factor concentrations

Figure 5-1 Local metabolic vasodilator hypothesis.

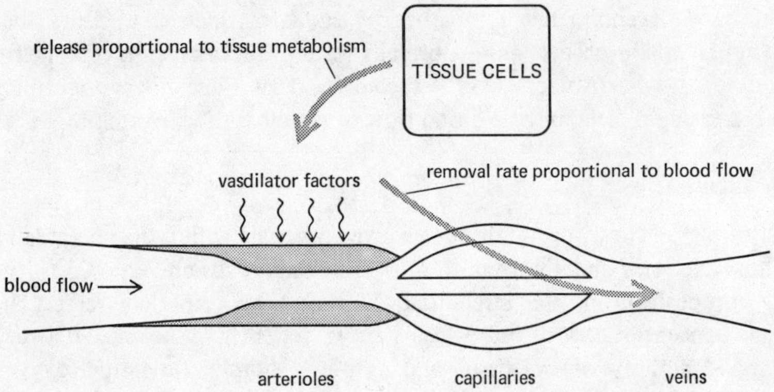

automatically build up and cause the arterioles to dilate. This, of course, causes blood flow to increase. The process continues until blood flow has risen sufficiently to appropriately match the tissue metabolic rate and prevent further accumulation of vasodilator factors. The same system also operates to reduce blood flow when it is higher than required by the tissue's metabolic activity, because this situation causes a reduction in the interstitial concentrations of metabolic vasodilator factors.

In organs with a highly variable metabolic rate, such as skeletal and cardiac muscle, the blood flow responds to and closely follows the tissue's metabolic rate. For example, skeletal muscle blood flow increases within seconds of the onset of muscle exercise and returns to control values shortly after exercise ceases. This phenomenon is known as *exercise* or *active hyperemia* (hyperemia means high flow). It should be clear how active hyperemia could result from the local metabolic vasodilator feedback on arteriolar smooth muscle.

Almost all organs respond to a period of reduced blood flow with a period of higher than normal flow once the flow restriction is removed. For example, flow through an extremity is higher than normal for a period after a tourniquet is removed from the extremity. This phenomenon is called *reactive* or *post-occlusion hyperemia.* In many (but perhaps not all) tissues, the cause of reactive hyperemia appears to be the interstitial accumulation of metabolic vasodilator factors during the period of reduced blood flow. The magnitude and duration of reactive hyperemia depend on the duration and severity of the occlusion as well as the concurrent metabolic rate of the tissue. These facts are all compatible with a local metabolic vasodilator mechanism for reactive hyperemia.

In addition to displaying reactive hyperemia, nearly all organs tend to keep their blood flow constant despite variations in arterial pressure; i.e., they *autoregulate* blood flow. If, for example, the blood pressure to a particular skeletal muscle is abruptly increased and kept high, the blood flow will initially increase but soon return toward the control level. The initial rise in flow with an increase in pressure is expected from the basic flow equation ($\dot{Q} = \Delta P/R$). The subsequent return of flow toward the control level despite a maintained increase in the driving pressure must involve an active increase in arteriolar tone and an increase in vascular resistance. Again, local metabolic influences on arteriolar tone provide an explanation for autoregulation in many organs because excess blood flow would tend to "wash out" vasodilator metabolites and cause vasoconstriction.

However, in certain organs such as the kidney, which autoregulates its blood flow very strongly yet does not demonstrate significant local metabolic control of arteriolar tone, mechanisms other than changes in vasodilator metabolite concentration may be responsible for the autoregulation phenomenon.

In addition to the metabolic theory of autoregulation stated above, there is a myogenic theory, in which it is assumed that arteriolar tone changes in response to mechanical (as well as chemical) influences. One possibility is that when arterioles are stretched by an increase in pressure, active changes in arteriolar tone occur so that arteriolar wall tension is held constant despite the change in pressure. According to the law of Laplace ($T = Pr$), arteriolar radius must decrease whenever the pressure inside arterioles increases if the arteriolar wall tension is to remain constant. Thus the active decrease in radius will increase vascular resistance and counteract the increase in flow induced by an increase in pressure.

Regardless of the mechanisms involved, all organs tend to autoregulate their blood flow to some degree.

Hormonal Influences on Arterioles

Hormonal influences on arterioles normally produce inconsequential effects on total peripheral resistance. It is true that measurable quantities of norepinephrine and epinephrine circulate in the blood. The primary source of these circulating catecholamines is the adrenal glands, where catecholamine release into the bloodstream is prompted by increased sympathetic nervous system activity. However, in all but perhaps the most severe circumstances, the levels of circulating catecholamines are too low to produce significant general vascular effects. For completeness, however, it should be recognized that the alpha-adrenergic receptors on smooth muscle could interact with circulating as well as neuronally released catecholamines. Vasoconstriction would result in either case. In addition, arteriolar smooth muscle has *beta-adrenergic receptors* similar, but not identical, to those found on cardiac muscle cells as well as *muscarinic cholinergic receptors* for acetylcholine. Both produce decreased arteriolar tone and vasodilation results when activated. Vascular beta-adrenergic receptors as well as cholinergic receptors (except those in arterioles of the external genitalia) are of more pharmacological than physiological interest.

Although many other substances that are found in circulating blood have vascular effects, the role of these substances in normal overall vascular control is minimal. Antidiuretic hormone (ADH), for example, was originally called vasopressin because in relatively high concentrations it was found to constrict isolated blood vessels. ADH actually has little general vascular effect at normal circulating levels, and it is now known that ADH plays its primary role in bodily fluid balance through its action on the renal collecting ducts.

In certain pathological situations, substances that normally have a negligible general vascular effect may have important cardiovascular influences. *Serotonin*, for example, is a substance normally confined inside platelets in the bloodstream. When there is tissue damage, however, serotonin released from platelets is thought to produce vasoconstriction and blood flow arrest in small vessels. *Histamine* is another substance released from traumatized tissue, and it causes arteriolar dilation and greatly increased capillary permeability. The local swelling often associated with tissue injury is due to transcapillary filtration of fluid, which is partly a consequence of arteriolar dilation and increased capillary permeability caused by histamine release.

Bradykinin is a potent arteriolar vasodilator substance released in association with increased activity of many secretory organs (e.g., salivary and sweat glands). Increased activity of a secretory organ requires an increase in organ blood flow, and bradykinin is thought to be important in the control of flow through such organs.

Angiotensin II is a circulating hormone whose normal function is thought to be the regulation of aldosterone release from the adrenal cortex as part of the system for controlling body sodium balance. It is, however, an extremely potent vasoconstrictor substance and is suspected of being instrumental in causing the generalized chronic increase in vascular resistance that accompanies most forms of hypertension. It should be emphasized that our knowledge of many pathological situations–including hypertension–is incomplete, and they may well involve vascular influences that are not yet recognized.

VENOUS CONTROL OF VASCULAR VOLUME

As emphasized in Chap. 4, the total amount of blood contained in an organ is determined largely by the diameter of its veins. Veins contain vascular smooth muscle that is influenced by many of the things that influence vascular smooth muscle of arterioles. Constriction of the veins (venoconstriction) is largely mediated through activity of the sympathetic nerves that innervate them. As in arterioles, these sympathetic nerves release norepinephrine, which interacts with alpha receptors and produces an increase in venous tone and a decrease in vessel diameter. There are, however, several functionally important differences between veins and arterioles. Compared to arterioles, veins normally have little intrinsic tone. Thus veins are normally in a vasodilated state and the peripheral venous blood pool is relatively full. One important consequence of the lack of intrinsic venous tone is that vasodilator metabolites that may accumulate in the tissue have little effect on veins.

Because of their thin walls, veins are much more susceptible to physical influences than are arterioles. The large effect of internal venous pressure on venous volume was discussed in Chap. 4. Often external compressional forces are an important determinant of venous volume. This is especially true of veins in skeletal muscle. Very high pressures are developed inside skeletal muscle tissue during contraction, which cause venous vessels to collapse. Because veins and venules have one-way valves, the blood displaced from veins during skeletal muscle contraction is forced in the forward direction toward the right heart. In fact, rhythmic skeletal muscle contractions can produce a considerable pumping action, often called the *skeletal muscle pump*, which helps return blood to the heart during exercise.

In summary, vessels are subject to a wide variety of influences and special influences often apply to particular organs. Certain general factors, however, dominate the control of the peripheral vasculature when it is viewed from the standpoint of overall cardiovascular system function, and these are summarized in Fig. 5-2. Intrinsic tone, local metabolic vasodilator factors, and sympathetic vasoconstrictor nerves acting through alpha receptors (α) are the major factors controlling arteriolar tone and therefore the blood flow rate through peripheral organs. Sympathetic vasoconstrictor nerves, internal pressure, and external compressional forces are the most important influences on venous diameter and therefore peripheral organ blood volume.

VASCULAR CONTROL IN SPECIFIC ORGANS

How the blood flow of a particular organ responds to changes in sympathetic vasoconstrictor fiber activity or changes in tissue metabolism depends, in large measure, on (1) the normal level of arteriolar tone and (2) what portion of the normal tone is intrinsic and what portion is neurogenic. Two examples are presented in Fig. 5-3 to illustrate this point.

In organs such as the brain, heart muscle, and skeletal muscle, the normal total arteriolar tone is high and composed largely of intrinsic tone, as illustrated in Fig. 5-3A. Usually the normal blood flow is not greatly in excess of that required to meet the normal metabolic demands of the tissue. As shown in Fig. 5-3A, changes in the activity of sympathetic vasoconstrictor fibers have much less effect on blood flow to these organs than do changes in their metabolic rate. Increasing sympathetic vasoconstrictor fiber activity does tend to reduce flow by causing vasoconstriction, but this causes a buildup of tissue metabolic vasodilators, which counteract the vasoconstriction and limit the

Figure 5-2 Major influences on blood vessels.

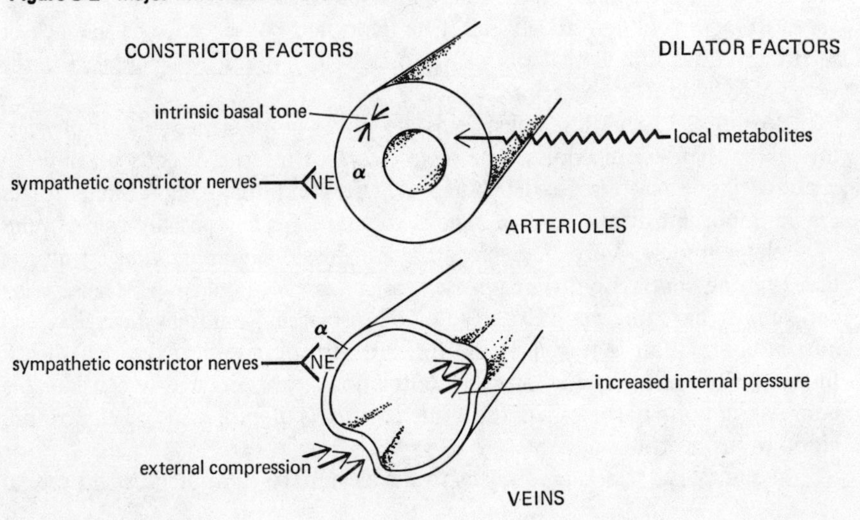

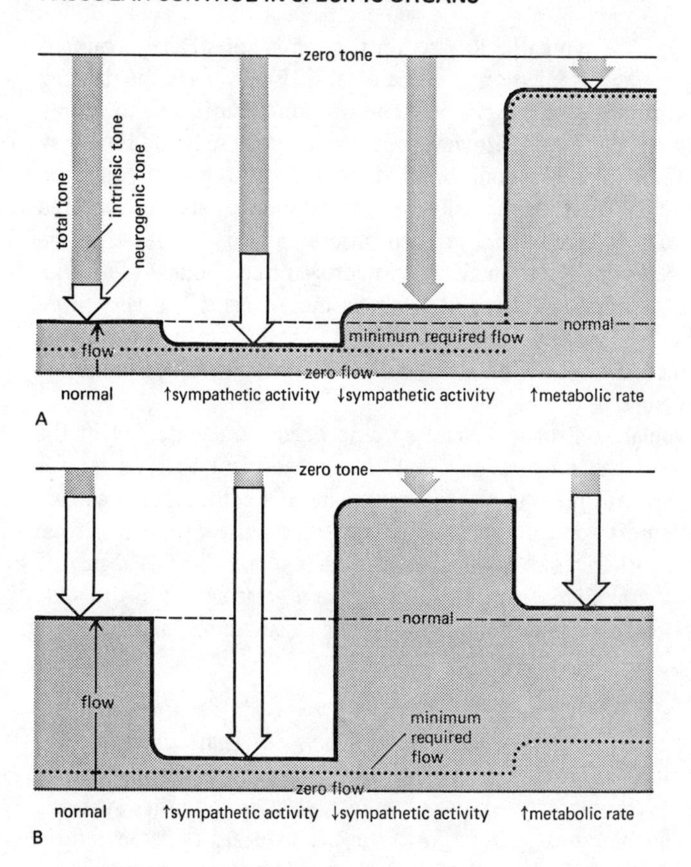

Figure 5-3 A. Flow control in an organ with normally *high intrinsic* arteriolar tone. B. Flow control in an organ with normally *high neurogenic* arteriolar tone.

extent of the flow reduction. Decreasing sympathetic vasoconstrictor fiber activity, on the other hand, can cause only a modest increase in flow in these organs since their *intrinsic* arteriolar tone is high. Increasing the tissue's metabolic rate and production of metabolically related vasodilator substances, however, can cause a large increase in flow by removing the normally high arteriolar tone.

A very different situation exists in the kidney, skin, and splanchnic organs, which normally have much more neurogenic than intrinsic arteriolar tone, as illustrated in Fig. 5-3B. The normal flow in these organs is usually well in excess of the minimum required for tissue metabolism and consequently the tissue concentrations of metabolically related vasodilator substances are very low. As indicated in Fig. 5-3B, increases in sympathetic vasoconstrictor fiber activity cause large reductions in flow to these organs. In part this is because the normal arteriolar tone is substantially less than the maximum possible arteriolar tone, and in part it is because sympathetic vasoconstriction is not strongly counteracted by local metabolic vasodilation in these organs. Usually, even the reduced blood flow accompanying sympathetic vasoconstriction is sufficient to supply the tissue's basic metabolic needs. However, such reductions in blood flow may well curtail whatever blood-reconditioning function the organ in question performs and, while tolerated temporarily, cannot last indefinitely. As indicated in Fig. 5-3B, decreasing sympathetic vasoconstrictor activity causes large increases in flow by removing the normally large neurogenic component of arteriolar tone. However, increasing metabolic rate in these organs has very little effect on blood flow since the normally high blood flow rate prevents metabolic vasodilator substances from accumulating to the tissue concentrations required to affect arteriolar tone.

Details of vascular control in several specific organs are presented in the remaining sections of this chapter. It should become evident that, as regards flow control, most organs may be placed somewhere in a continuous spectrum that ranges from almost total dominance by local mechanisms to almost total dominance by sympathetic nerves. Organs in which blood flow is regulated predominantly by sympathetic nerves participate to a great extent in the wide variety of cardiovascular reflex responses that will be discussed in Chap. 7.

Coronary Blood Flow

The major right and left coronary arteries that serve the heart tissue are the first vessels to branch off the aorta. Thus the driving force for myocardial blood flow is the systemic arterial pressure, just as it is for other systemic organs. Most of the blood that flows through the myocardial tissue returns to the right atrium by way of a large cardiac vein called the coronary sinus.

Local Metabolic Control As emphasized before, coronary blood flow is controlled primarily by local metabolic mechanisms and thus it responds rapidly and accurately to changes in myocardial oxygen consumption. In a resting individual, the oxygen concentration in coronary sinus blood is very low ($\approx$ 25 percent of the arterial blood oxygen concentration). The Fick principle makes it clear, therefore, that myocardial oxygen consumption can increase significantly only when coronary blood flow also increases significantly. In fact, coronary blood flow normally follows myocardial oxygen consumption so closely that the oxygen levels in coronary sinus blood are essentially constant regardless of the rate of myocardial oxygen consumption.

The issue of which metabolic vasodilator factor(s) play the dominant role in modulating the tone of coronary arterioles is unresolved at present. Many believe that adenosine, released from myocardial muscle cells in response to lack of oxygen, may be the most important local coronary metabolic vasodilator influence. Regardless of the specific details, myocardial oxygen consumption is the most important influence on coronary blood flow.

Systolic Compression Large forces and/or pressures are generated *within* the myocardial *tissue* during cardiac muscle contraction. Such intramyocardial forces press on the outside of coronary vessels and cause them to collapse during systole. Because of this *systolic compression* and the associated collapse of coronary vessels, coronary vascular resistance is very high during systole. The result, at least for much of the left ventricular myocardium, is that coronary flow is lower during systole than during diastole, even though systemic arterial pressure is highest during systole. This is illustrated in the left coronary artery flow trace shown in Fig. 5-4. Systolic compression has much less effect on flow through the right ventricular myocardium, as is evident from the right coronary artery flow trace in Fig. 5-4. This is because the peak systolic intraventricular pressure is much lower for the right heart than the left, and the systolic compressional forces in the right ventricular wall are correspondingly less than those in the left ventricular wall.

Systolic compressional forces on coronary vessels happen to be greater in the endocardial (inside) layers of the left ventricular wall than in the epicardial layers.[1] Thus the flow to the endocardial layers of the left ventricle is impeded

[1] Consider that the endocardial surface of the left ventricle is exposed to intraventricular pressure (≈ 120 mmHg during systole), while the epicardial surface is exposed only to intrathoracic pressure (≈ 0 mmHg).

Figure 5-4 Phasic flows in the left and right coronary arteries in relation to aortic and left ventricular pressures.

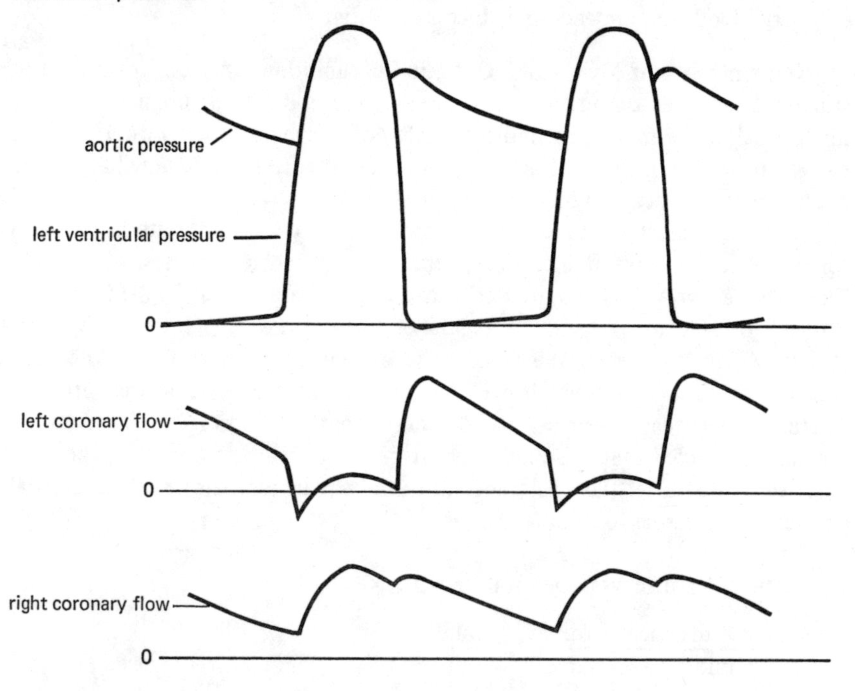

more than the flow to epicardial layers by systolic compression. Normally the endocardial region of the myocardium can make up for the lack of flow during systole by a high flow in the diastolic interval. However, when coronary blood flow is limited—for example, by coronary disease and stenosis—the endocardial layers of the left ventricle are often the first region of the heart to have difficulty maintaining a flow sufficient for its metabolic needs. *Myocardial infarcts* (areas of tissue killed by lack of blood flow) occur most frequently in the endocardial layers of the left ventricle.

Neural Influences on Coronary Flow Coronary arterioles are densely innervated with sympathetic vasoconstrictor fibers, yet when the activity of the sympathetic nervous system increases, the coronary arterioles normally vasodilate rather than vasoconstrict. This is because an increase in sympathetic tone increases myocardial oxygen consumption by increasing heart rate and contractility. The increased local metabolic vasodilator influence apparently overwhelms any concurrent vasoconstrictor influence due to an increase in the activity of sympathetic vasoconstrictor fibers that terminate on coronary arterioles. Consequently, coronary sympathetic vasoconstrictor fibers are thought to have very little influence on blood flow to the normal heart. Whether these coronary vasoconstrictor fibers might be functionally important in certain pathological situations is an open question.

As mentioned previously, coronary arterioles also receive parasympathetic vasodilator fiber innervation. However, their role in the normal control of coronary blood flow appears to be inconsequential.

Determinants of Myocardial Oxygen Consumption In many pathological situations, such as coronary artery disease, the oxygen requirements of the myocardial tissue may exceed the capacity of coronary blood flow to deliver oxygen to the heart tissue. It is important, therefore, to understand what factors determine the myocardial oxygen consumption rate.

The heart must do a measurable amount of work each minute in order to force the cardiac output into the arterial system, where the pressure is high. The external work (W) required to move a given volume of fluid (ΔV) from a region of zero pressure into a region of higher pressure (P) is given by $W = \int_0^{\Delta V} P\,dV$. If we consider the work done by the left heart in 1 min, then ΔV is equal to cardiac output. If we also neglect the fact that arterial pressure fluctuates with each heartbeat and assume that all portions of each stroke volume are ejected against some constant mean arterial pressure ($\overline{P}_A$), then the external work the left ventricle does in 1 min may be estimated as mean arterial pressure times cardiac output (CO):

$$\text{External cardiac work per minute} = \overline{P}_A \times \text{CO}$$

$$\frac{\text{Force} \times \text{distance}}{\text{min}} = \frac{\text{force}}{\text{area}} \times \frac{\text{volume}}{\text{min}}$$

We see from this equation that doubling either the cardiac output or the mean arterial pressure would double the external work the left ventricle does each minute. Increasing the work the heart must perform per minute increases the myocardial oxygen consumption rate. However, it is a fact that increasing cardiac work by increasing only arterial pressure requires a greater increase in myocardial oxygen consumption than does the same increase in cardiac work caused by increasing only the cardiac output. Thus *"pressure work" is more costly to the heart than "flow work."* One detrimental feature (certainly not the only one) of hypertension (high blood pressure), for example, is that it significantly increases the heart's oxygen requirements.

On close inspection, myocardial oxygen consumption is rather poorly predicted by the external work of the heart. Nearly one-fourth of the oxygen consumed by the heart appears to supply energy for cellular processes other than contraction, e.g., energy-dependent ion pumping. Moreover, much more adenosine triphosphate (ATP) is required by the contractile proteins in generating wall tension and ventricular pressure during isovolumetric contraction than is required during the phase of muscle fiber shortening and cardiac ejection, when the external work for each heartbeat is actually performed. Again, developing pressure requires more energy than producing flow does.

Achieving a given level of cardiac output with a high heart rate and low stroke volume requires more myocardial oxygen consumption than producing the same cardiac output with a lower heart rate and higher stroke volume. The reason is that, if all other factors are constant, the more times per minute that the heart must develop intraventricular pressure, the greater is the energy that is required.

It actually appears that it is not so much the peak systolic intraventricular pressure as the peak systolic ventricular wall tension that is best correlated with myocardial oxygen consumption. This is especially important in pathological situations such as heart failure where the heart is dilated (ventricular diameter larger than normal).[2] Dilated hearts must develop a greater wall tension than normal hearts to produce a normal intraventricular pressure ($T = Pr$). Thus a dilated heart is likely to have a higher than normal oxygen consumption rate even when cardiac output and arterial pressure are normal.

If all other factors remain constant, increasing myocardial contractility increases myocardial oxygen consumption. In part this is because increasing the contractility of cardiac muscle cells increases the rate at which they develop isometric tension, and this requires more rapid cross-bridge cycling and thus ATP splitting.

In summary, the factors that determine myocardial oxygen consumption are complex and, as yet, incompletely defined. Myocardial energy requirements are increased whenever the external work of the heart is increased by an increase

[2] Cardiac dilation is not be confused with cardiac hypertrophy, which denotes higher than normal myocardial muscle mass and ventricular wall thickness.

in either cardiac output or arterial pressure. However, changes in arterial pressure have a greater influence on myocardial oxygen consumption than do changes in cardiac output. Furthermore, even when arterial pressure and cardiac output are normal, myocardial oxygen consumption may be elevated by increased heart rate, increased contractility, and/or cardiac dilation (wherein normal intraventricular pressure development requires higher than normal wall tension).

Coronary Artery Disease Whenever coronary blood flow falls below that required to meet the metabolic needs of the heart, the myocardium is said to be *ischemic.* Myocardial ischemia not only impairs the pumping ability of the heart, but also produces intense chest pain called *angina pectoris.* The most common cause of myocardial ischemia and angina is atherosclerotic disease of large coronary arteries. In atherosclerotic disease, localized lipid deposits called plaques develop within the arterial walls. With severe disease these plaques may become so large that they physically narrow the lumen of arteries (producing a stenosis) and thus greatly and permanently increase the normally low vascular resistance. This extra resistance, of course, adds to the resistance of other coronary vascular segments and tends to reduce coronary flow. If the coronary artery stenosis is not too severe, local metabolic vasodilator mechanisms may reduce arteriolar resistance sufficiently to compensate for the abnormally large arterial resistance. Thus an individual with coronary artery disease may have perfectly normal coronary blood flow when resting. A coronary artery stenosis of any significance will, however, limit the extent to which coronary flow can increase above its resting value by reducing maximum achievable coronary flow. This occurs because, even with very low *arteriolar* resistance the overall vascular resistance of the coronary vascular bed is high if *arterial* resistance is high. Anginal pain is often absent in individuals with coronary artery disease when they are resting but is induced during physical exertion or emotional excitement. Both exertion and excitement elicit an increase in sympathetic tone and this increases myocardial oxygen consumption. Myocardial ischemia and chest pain may result in both situations if coronary blood flow cannot keep pace with the increase in myocardial metabolism.

Myocardial ischemia is a dangerous situation because ischemic muscle cells are electrically irritable and unstable. During ischemia the normal cardiac electrical excitation pathways may be upset and often ectopic foci develop. Thus the danger of fibrillation is great during ischemic episodes.

The pharmacological treatment of coronary artery disease follows either or both of two paths. First, coronary vasodilator drugs such as nitroglycerin may be used to increase coronary blood flow. Second, measures such as administering beta-receptor blocking agents may be taken to block the effects of the cardiac sympathetic nerves on heart rate and contractility. These agents limit myocardial oxygen consumption and prevent it from increasing above the level that the compromised coronary blood flow can sustain. Often surgical interventions are used in an attempt to bypass the stenotic coronary artery segments with parallel low-resistance pathways formed from either natural or artificial vessels.

Skeletal Muscle Blood Flow

Collectively, the skeletal muscles constitute 40 to 45 percent of body weight—more than any other single body organ. Even at rest, about 15 percent of the cardiac output goes to skeletal muscle, and during strenuous exercise skeletal muscle may receive up to 70 percent of the cardiac output. Thus skeletal muscle blood flow is an important factor in overall cardiovascular hemodynamics.

Because of the high level of intrinsic tone of the resistance vessels in resting skeletal muscles, the blood flow per gram of tissue is quite low when compared to other organs such as the kidneys. However, resting skeletal muscle blood flow is still substantially above that required to sustain its metabolic needs. Resting skeletal muscles normally extract only 25 to 30 percent of the oxygen delivered to them in arterial blood. Changes in the activity of sympathetic vasoconstrictor fibers can alter the resting blood flow. For example, maximum sympathetic discharge rates can decrease blood flow in a resting muscle to less than one-fourth its normal value, and conversely, if all neurogenic tone is removed, resting skeletal muscle blood flow may double. This is, however, a modest increase in flow compared to the 20-fold increase in flow that can occur in an exercising skeletal muscle.

A particularly important characteristic of skeletal muscle is its very wide range of metabolic rates. During heavy exercise the oxygen consumption rate of skeletal muscle tissue can reach the high values typical of the myocardium. An exercising skeletal muscle can utilize 80 to 90 percent of the oxygen carried to it in the arterial blood. In most respects, the factors that control blood flow to exercising muscle are similar to those that control coronary blood flow. Local metabolic control of arteriolar tone is very strong in exercising skeletal muscle, and muscle oxygen consumption is the most important determinant of blood flow in exercising skeletal muscle. Sympathetic nerves have relatively little effect on the blood flow of an exercising skeletal muscle. The control of skeletal muscle blood flow during rest and exercise generally follows the patterns illustrated in Fig. 5-3A.

As in the heart, muscle contraction produces large compressional forces within the tissue, which can collapse vessels and impede blood flow. Strong, sustained (tetanic) skeletal muscle contractions may actually stop muscle blood flow. About 10 percent of the total blood volume is normally contained within the veins of skeletal muscle, and during rhythmic exercise the muscle pump is very effective in displacing blood from skeletal muscle veins. Blood displaced from skeletal muscle into the central venous pool is an important factor in the hemodynamics of strenuous whole body exercise.

The veins in skeletal muscle are rather sparsely innervated with sympathetic vasoconstrictor fibers, and the rather small volume of blood that can be mobilized from skeletal muscle by sympathetic nerve activation is probably not of much significance to total body hemodynamics. This is in sharp contrast to the large displacement of blood from exercising muscle by the muscle pump mechanism.

Cerebral Blood Flow

Adequate cerebral blood flow is of paramount importance for survival because unconsciousness occurs very rapidly after an interruption in flow. One rule of overall cardiovascular system function is that, in *all* situations, measures are taken that are appropriate to preserve adequate blood flow to the brain.

The brain tissue itself has a nearly constant rate of metabolism that, on a per gram basis, is nearly as high as that of myocardial tissue. Cerebral blood flow is determined almost entirely by local mechanisms. Although cerebral arterioles receive both sympathetic vasoconstrictor and parasympathetic vasodilator fiber innervation, cerebral blood flow is influenced very little by changes in the activity of either.

Presumably because the metabolic rate of brain tissue shows little variation, total brain blood flow is remarkably constant over nearly all situations. Flow through the cerebrum is autoregulated very strongly and is little affected by changes in arterial pressure unless it falls below about 60 mmHg. When arterial pressure decreases below 60 mmHg, brain blood flow decreases proportionately.

Cerebral blood flow does increase whenever the partial pressure of carbon dioxide (P_{CO_2}) is raised above normal in the arterial blood. Conversely, cerebral blood flow decreases whenever arterial blood P_{CO_2} falls below normal. It appears most likely that cerebral arterioles respond not to changes in P_{CO_2} but to changes in the H^+ concentration (i.e., pH) caused by changes in P_{CO_2}. Cerebral arterioles also vasodilate whenever the partial pressure of oxygen (P_{O_2}) in arterial blood falls significantly below normal values. The precise mechanisms and key metabolic factors involved in the local metabolic control of cerebral blood flow, however, remain obscure.

The cerebral circulation is distinguished by two important anatomic features. First, brain capillaries are considerably less porous than those in other organs and form what is called the *blood-brain barrier.* Transcapillary movement of even small particles is restricted so much that the cerebral interstitial fluid constitutes a distinct fluid compartment called the *cerebrospinal fluid.* Unlike the interstitial fluid of other tissues, cerebrospinal fluid differs in ionic composition from blood plasma. Second, the brain tissue volume is fixed by the rigid cranium. Significant alterations in brain blood volume simply cannot occur.

Splanchnic Blood Flow

A number of abdominal organs, including the gastrointestinal tract, spleen, pancreas, and liver, are collectively supplied with what is called the splanchnic blood flow. Splanchnic blood flow is supplied to these abdominal organs through many arteries, but it all ultimately passes through the liver and returns to the inferior vena cava through the hepatic veins.

The organs of the splanchnic region receive about 25 percent of the resting cardiac output and moreover contain more than 20 percent of the circulating blood volume. Thus adjustments in either the blood flow or the blood volume of this region have extremely important effects on the cardiovascular system.

There is a great diversity of functions among individual organs and even regions within organs in the splanchnic region. Blood flow is required to support secretory and absorptive processes as well as muscular contractions of the gastrointestinal tract. The mechanisms of vascular control in specific areas of the splanchnic region are not well understood but are likely to be quite varied. Nonetheless, since most of the splanchnic organs are involved in the digestion and absorption of food from the gastrointestinal tract, splanchnic blood flow increases after food ingestion. A large meal can elicit a 30 to 100 percent increase in splanchnic flow, but individual organs in the splanchnic region probably have higher percentage increases in flow at certain times because they are involved sequentially in the digestion-absorption process.

Collectively, the splanchnic organs have a relatively high blood flow and extract only 15 to 20 percent of the oxygen delivered to them in the arterial blood. In general, the situation of Fig. 5-3B applies to the splanchnic bed and the sympathetic nerves play a significant role in vascular control. Maximal activation of sympathetic vasoconstrictor nerves can produce an 80 percent reduction in flow to the splanchnic region and also cause a large shift of blood from the splanchnic organs to the central venous pool.

Renal Blood Flow

The kidneys normally receive approximately 20 percent of the cardiac output of a resting individual, and since this can be reduced to practically zero, the control of renal blood flow is important to overall cardiovascular control. However, because the kidneys are such small organs, changes in renal blood volume are inconsequential to overall cardiovascular hemodynamics.

Although the renal vascular bed is specialized in many ways that are important to renal function (e.g., two distinct capillary beds arranged in series), renal blood flow follows the patterns of adjustment shown in Fig. 5-3B well. Renal resistance vessels are normally under the influence of substantial tonic sympathetic vasoconstrictor fiber activity. Increases in this activity can markedly reduce total renal blood flow by increasing the neurogenic tone of renal resistance vessels. In fact, extreme situations involving intense and prolonged sympathetic vasoconstrictor activity can lead to renal failure.

As discussed earlier, the renal vascular resistance adjusts to keep renal blood flow nearly constant over a wide range of arterial pressures; i.e., the kidneys autoregulate strongly. Renal autoregulation is, however, probably of more importance to renal function than overall cardiovascular hemodynamics. Renal function is itself of paramount importance to overall cardiovascular function, as will be described in Chap. 7.

Cutaneous Blood Flow

The metabolic activity of body cells produces heat, which must be lost in order for the body temperature to remain constant. The skin is the primary site of exchange of body heat with the external environment. Alterations in cutaneous blood flow in response to various metabolic states and environmental conditions provide the primary mechanism responsible for temperature homeostasis (other mechanisms such as shivering, sweating, and panting also participate in body temperature regulation under more extreme conditions).

Cutaneous blood flow, which is about 6 percent of the resting cardiac output, can decrease to about one-twentieth of its normal value when heat is to be retained (e.g., in a cold environment, during the development stages of a fever). On the other hand, cutaneous blood flow can increase up to seven times its normal value when heat is to be lost (e.g., in a hot environment, accompanying a high metabolic rate, after a fever breaks).

The actual anatomic interconnections between microvessels in the skin are highly specialized and extremely complex. An extensive system of interconnected veins called the *venous plexus* normally contains the largest fraction of the cutaneous blood volume, which, in individuals with lightly pigmented skin, gives the skin a reddish hue. To a large extent, heat transfer from the blood takes place across the large surface area of the venous plexus. The venous plexus is richly innervated with sympathetic vasoconstrictor nerves. When these fibers are activated, blood is displaced from the venous plexus, and this helps reduce heat loss and also lightens the skin color. Since the skin is one of the largest body organs, venous constriction can shift a considerable amount of blood into the central venous pool.

Cutaneous resistance vessels are also richly innervated with sympathetic vasoconstrictor nerves, and since these fibers have a normal tonic activity, cutaneous resistance vessels normally have a high degree of neurogenic tone. In general, cutaneous blood flow follows the response patterns shown in Fig. 5-3B. During sweating, however, cutaneous blood flow rises considerably above that produced by removing sympathetic vasoconstrictor tone. It is postulated that additional cutaneous vasodilation occurs during sweating because the powerful vasodilator substance bradykinin is produced locally in association with sweat gland activity. Also, skin cooling causes cutaneous vasoconstriction and skin heating causes cutaneous vasodilation by some unknown but local mechanism.

Pulmonary Blood Flow

The rate of blood flow through the lungs is necessarily equal to cardiac output in all circumstances. When cardiac output increases threefold during exercise, for example, pulmonary blood flow must also increase threefold. Whereas the flow through a systemic organ is determined by its vascular resistance ($\dot{Q} = \Delta P/R$), the blood flow rate through the lungs is determined simply by the cardiac output ($\dot{Q} = \text{CO}$). Pulmonary vessels do, however, offer some vascular resistance. Although the level of pulmonary vascular resistance does not usually influence the pulmonary flow rate, it is important because it is one of the determinants of pulmonary arterial pressure ($\Delta P = \dot{Q}R$). Recall that mean *pulmonary* arterial pressure is about 13 mmHg, whereas mean *systemic* arterial pressure is about 100 mmHg. The reason for the difference in pulmonary and systemic arterial pressures is not that the right heart is weaker than the left heart but rather that pulmonary vascular resistance is inherently much lower than systemic total peripheral resistance. The pulmonary bed has a low resistance because it has relatively large vessels throughout.

A very important distinction between the systemic and pulmonary arteries and arterioles is that the pulmonary vessels are much more compliant. When pulmonary arterial pressure increases, the pulmonary arteries and arterioles become larger in diameter. Thus an increase in pulmonary arterial pressure *decreases* pulmonary vascular resistance. Pulmonary resistance vessels are so compliant that a 5-mmHg rise in pulmonary arterial pressure can lower pulmonary vascular resistance to one-fourth its initial value. Thus a fourfold increase in cardiac output causes only about a 5-mmHg rise in pulmonary arterial pressure rather than the fourfold increase (to ≈ 50 mmHg) that would occur with fixed pulmonary vascular resistance ($\Delta P = \dot{Q}R$).

Pulmonary vessels in general are much less muscular than systemic vessels, and active control of pulmonary vascular resistance or volume is quite limited. Changes in pulmonary vascular resistance and blood volume occur primarily for passive and physical reasons. An important exception is the phenomenon of hypoxic (low oxygen) vasoconstriction of pulmonary arterioles, which causes blood flow to be diverted away from the areas of the lung tissue that for some reason (e.g., a blocked airway) are not being ventilated.

An important consequence of the low mean pulmonary arterial pressure is the low pulmonary capillary hydrostatic pressure of about 8 mmHg (compared to 25 mmHg in systemic capillaries). Because of the low pulmonary capillary hydrostatic pressure, the forces for transcapillary fluid movement are not normally in balance in the lung. Since the capillary oncotic, tissue hydrostatic, and tissue oncotic pressures have their usual values in the lung, there is normally a large (≈ 17 mmHg) pressure imbalance for transcapillary fluid reabsorption in the lung. The constant tendency for fluid reabsorption ensures that the lung tissue normally stays dry. Moreover, a large rise in pulmonary capillary hydrostatic pressure is required before pulmonary edema formation begins.

Study questions: **24** to **30**

Chapter 6

Venous Return and Cardiac Output

OBJECTIVES

The student understands how venous return, cardiac output, and central venous pressure are interrelated:

1 Defines venous return and explains how it is distinguished from cardiac output.
2 States the reason why cardiac output and venous return must be equal in the steady state.
3 Lists the factors that control venous return.
4 Describes the relationship between venous return and central venous pressure and draws the normal venous return curve.
5 Defines peripheral venous pressure.
6 Lists the factors that determine peripheral venous pressure.
7 Predicts the shifts in the venous return curve that occur with altered blood volume and altered venous tone.
8 Draws the normal venous return and cardiac output curves on a graph and describes the significance of the point of curve intersection.

9 Predicts how normal venous return, cardiac output, and central venous pressure will be altered with any given combination of changes in cardiac sympathetic tone, peripheral venous sympathetic tone, or circulating blood volume.

Any adjustment made by a single component in the cardiovascular system produces hemodynamic alterations throughout the system. For example, an increase in peripheral venous tone usually results in increased cardiac output. In this chapter we will describe the interactions that occur between the heart and the peripheral vasculature at the connection between them on the venous side.

Recall that we have described a space, called the central venous pool, that corresponds loosely to the volume enclosed by the right atrium and the great veins in the thorax. Blood leaves the central venous pool by entering the right ventricle. In any steady situation the rate at which blood enters the heart must be identical to cardiac output. Therefore, blood *leaves* the central venous pool at a rate equal to the cardiac output. *Venous return*, on the other hand, is by definition the rate at which blood returns to the thorax from the peripheral vascular beds and thus is the rate at which blood flows *into* the central venous pool. The important distinction between venous return and cardiac output is illustrated in Fig. 6-1.

In any stable situation, venous return must precisely equal cardiac output or blood would gradually accumulate in the central venous pool or the peripheral vasculature. However, there can be, and often are, temporary differences between cardiac output and venous return. Whenever such differences exist, the volume of the central venous pool must be changing. Since the central venous pool is enclosed by elastic tissues, any change in central venous volume produces a corresponding change in central venous pressure. We discussed in Chap. 3 how any change in central venous pressure changes cardiac output (Starling's law). As described below, alterations in central venous pressure also

Figure 6-1 Distinction between cardiac output and venous return.

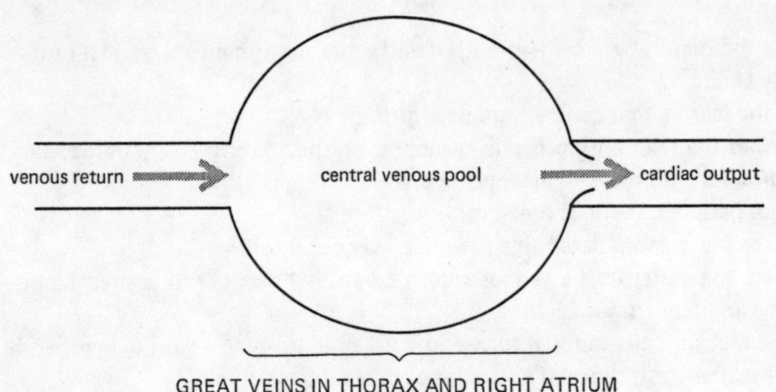

change venous return. Thus whenever an influence acts on the heart to change cardiac output, a change in central venous pressure is automatically produced that causes an appropriate change in venous return. Conversely, whenever venous return is altered by a peripheral vascular influence, a change in central venous pressure is automatically produced that causes an appropriate adjustment in cardiac output. To appreciate these concepts more fully, we must first understand how central venous pressure influences venous return.

VENOUS RETURN CURVE

The important factors involved in the process of venous return can be summarized as shown in Fig. 6-2A. Basically, blood flows from the peripheral venous pool to the central venous pool through some small (but not negligible) venous vascular resistance. Anatomically the peripheral venous pool is scattered throughout the systemic organs, but functionally it can be viewed as a single vascular space that has a particular pressure (P_{PV}) at any instant of time. The blood flow rate between the peripheral venous pool and the central venous pool is governed by the basic flow equation ($\dot{Q} = \Delta P/R$). In the example of Fig. 6-2, peripheral venous pressure is assumed to be 7 mmHg. Thus there will be no venous return when the central venous pressure (P_{CV}) is also 7 mmHg. This situation is represented in the graph of Fig. 6-2B as the intersection of the venous return curve with the central venous pressure axis at 7 mmHg. If the peripheral venous pressure remains at 7 mmHg, then decreasing central venous pressure will increase the pressure drop across the venous resistance and consequently cause an increase in venous return. All this is summarized by the *venous return curve*, which shows how venous return increases as central venous pressure drops.[1] If central venous pressure reaches very low values and falls below the intrathoracic pressure, the veins in the thorax collapse and tend to limit venous return. In the example of Fig. 6-2, intrathoracic pressure is taken to be 0 mmHg and the flat portion of the venous return curve indicates that lowering central venous pressure below 0 mmHg produces no additional increase in venous return.

Just as a cardiac function curve shows how central venous pressure influences cardiac output, a venous return curve shows how central venous pressure influences venous return, if other factors remain constant.

[1] The slope of the venous return curve is determined by the value of the venous vascular resistance. Lowering the venous vascular resistance would tend to raise the venous return curve and make it steeper because more venous return would result for a given difference between P_{PV} and P_{CV}. However, if P_{PV} is 7 mmHg, venous return will be zero when $P_{CV} = 7$ mmHg at any level of venous vascular resistance ($\dot{Q} = \Delta P/R$). We have chosen to ignore the complicating issue of changes in venous vascular resistance because they do not affect the general conclusions to be drawn from the discussion of venous return curves.

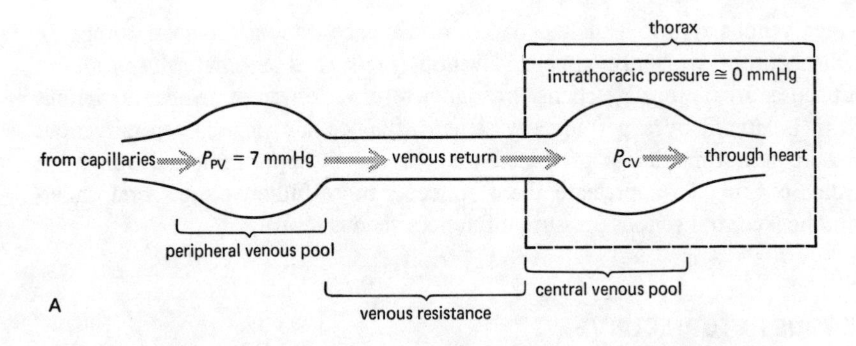

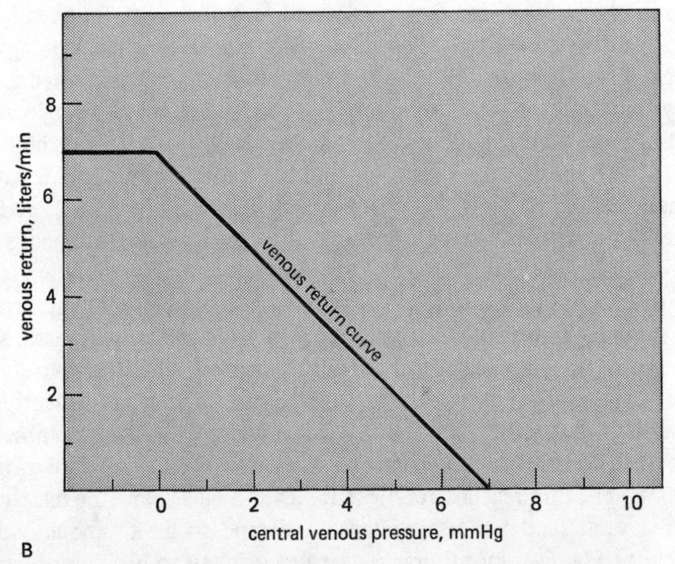

Figure 6-2 A. Factors influencing venous return. B. The venous return curve.

INFLUENCE OF PERIPHERAL VENOUS PRESSURE ON VENOUS RETURN

As can be deduced from Fig. 6-2A, it is the pressure difference between the peripheral and central venous pools that determines venous return, and an increase in peripheral venous pressure is just as effective in increasing venous return as a drop in central venous pressure.

The two ways in which peripheral venous pressure can change were discussed in Chap. 4. First, because veins are elastic vessels, changes in the volume of blood contained within the peripheral veins alter the peripheral venous pressure. Moreover, since the veins are much more compliant than any other vascular segment, changes in circulating blood volume produce larger changes in the volume of blood in the veins than in any other vascular segment. For example, blood loss by hemorrhage or loss of body fluids through severe sweating, vomiting, or diarrhea will decrease circulating blood volume and significantly decrease the volume of blood contained in the veins. On the other hand, transfusion, fluid retention by the kidney, or transcapillary fluid reabsorption will increase circulating blood volume and increase venous blood volume. Thus whenever circulating blood volume increases, peripheral venous pressure increases.

Recall from Chap. 4 that the second way that peripheral venous pressure can be altered is through changes in venous tone produced by increasing or decreasing the activity of sympathetic vasoconstrictor nerves supplying the venous smooth muscle. Peripheral venous pressure increases whenever the activity of sympathetic vasoconstrictor fibers to veins increases. In addition, an increase in any force compressing veins from the outside has the same effect on the pressure inside veins as an increase in venous tone. Thus such things as muscle exercise and wearing elastic stockings tend to increase peripheral venous pressure.

Whenever peripheral venous pressure is altered, the relationship between central venous pressure and venous return is also altered. For example, whenever peripheral venous pressure is increased by increases in blood volume or by sympathetic stimulation, the venous return curve shifts upward and to the right, as shown in Fig. 6-3. This relationship can be most easily understood by focusing first on the central venous pressure at which there will be no venous return. When peripheral venous pressure is 7 mmHg, venous return is zero

Figure 6-3 Effect of changes in blood volume and venous tone on venous return curves.

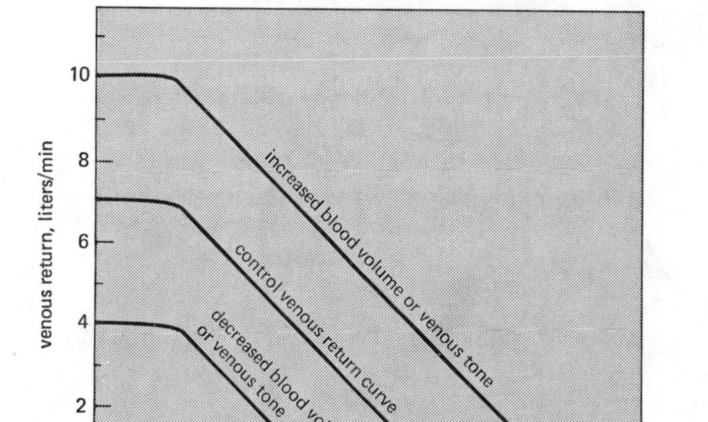

when central venous pressure is 7 mmHg. When peripheral venous pressure is increased to 10 mmHg, considerable venous return occurs with a central venous pressure of 7 mmHg, and venous return stops only when central venous pressure is raised to 10 mmHg. Thus, increasing peripheral venous pressure shifts the whole venous return curve to the right. By similar logic, decreased peripheral venous pressure caused by blood loss or decreased sympathetic vasoconstriction of peripheral veins shifts the venous return curve to the left, as indicated in Fig. 6-3.

DETERMINATION OF CARDIAC OUTPUT AND VENOUS RETURN BY CENTRAL VENOUS PRESSURE

The significance of the fact that central venous pressure simultaneously affects both cardiac output and venous return can be best seen by plotting the cardiac output curve (Starling's law) and the venous return curve on the same graph, as in Fig. 6-4.

Note that in Fig. 6-4, cardiac output and venous return are equal (at 5 liters/min) *only* when the central venous pressure is 2 mmHg. If central venous pressure were to decrease to 0 mmHg for any reason, cardiac output would fall (to 2 liters/min) and venous return would increase (to 7 liters/min). With a venous return of 7 liters/min and a cardiac output of 2 liters/min, the volume of the central venous pool would necessarily be increasing and this would

Figure 6-4 Interaction of cardiac output and venous return through central venous pressure.

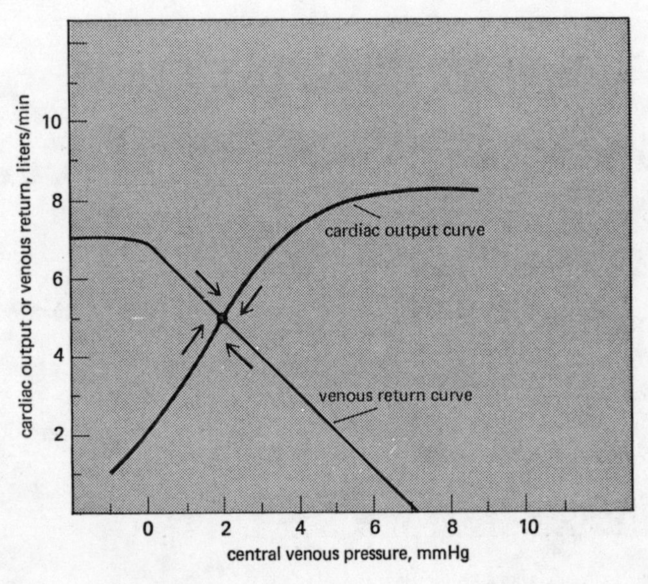

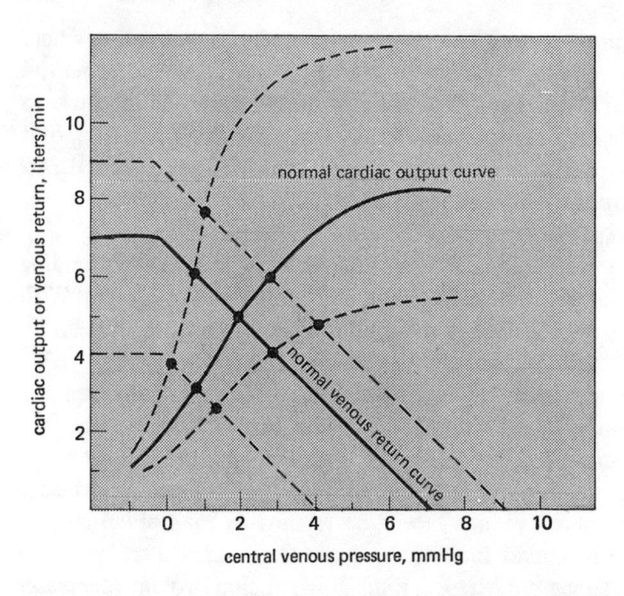

Figure 6-5 Families of cardiac output and venous return curves. Intersection points indicate equilibrium values for cardiac output, venous return, and central venous pressure.

produce a progressively increasing central venous pressure. In this manner, central venous pressure would return to the original level (2 mmHg) in a very short time. On the other hand, if central venous pressure were to increase from 2 to 4 mmHg for any reason, venous return would decrease (to 3 liters/min) and cardiac output would increase (to 7 liters/min). This would quickly decrease the volume of blood in the central venous pool and the central venous pressure would soon fall back to the original level. The cardiovascular system always and automatically adjusts to operate at the point where the cardiac output and venous return curves intersect. *Central venous pressure is always inherently driven to the equilibrium value that makes cardiac output and venous return equal. Cardiac output (and venous return) always stabilizes at the level where the cardiac output and venous return curves intersect.*

In order to fulfill its homeostatic role in the body, the cardiovascular system must be able to alter its cardiac output. Recall from Chap. 3 that cardiac output is affected by more than just filling pressure and that at any moment the heart may be operating on any one of a number of cardiac output curves, depending on the existing level of cardiac sympathetic tone (Fig. 3-10). The family of possible cardiac output curves may be plotted along with the family of possible venous return curves, as shown in Fig. 6-5. At a particular moment the existing influences on the heart dictate the particular cardiac output curve on which it is working, and similarly, the existing influences on peripheral venous pressure dictate the particular venous return curve that applies. Thus the influences on the heart and on the peripheral vasculature determine where the cardiac output and venous return curves intersect and thus what the central venous pressure and cardiac output (and venous return) are at equilibrium. In the intact cardiovascular system, cardiac output can rise only when the point of intersection of the cardiac output and venous return curves is raised. *All changes in cardiac output are caused by a shift in the cardiac output curve, a shift in the venous return curve, or both.*

The cardiac output and venous return curves are useful for understanding the complex interactions that occur in the intact cardiovascular system. With the help of Fig. 6-6, let us consider, for example, what happens to the cardiovascular system when there is a significant loss of blood (hemorrhage). We assume that before the hemorrhage sympathetic tone to the heart and peripheral vessels is normal, as is the blood volume. Therefore cardiac output is related to central venous pressure as indicated by the "normal" cardiac output curve in Fig. 6-4. In addition, venous return is determined by central venous pressure as indicated by the "normal" venous return curve shown. The normal cardiac output and venous return curves intersect at point A, so cardiac output is 5 liters/min and central venous pressure is 2 mmHg. When blood volume decreases due to hemorrhage, the peripheral venous pressure falls and the venous return curve is shifted to the left. In the absence of any cardiovascular responses, the cardiovascular system must switch its operation to point B because this is now the point at which the cardiac output and venous return curves intersect. Note that at the moment after the blood loss shifts the venous return curve, venous

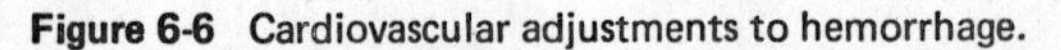

Figure 6-6 Cardiovascular adjustments to hemorrhage.

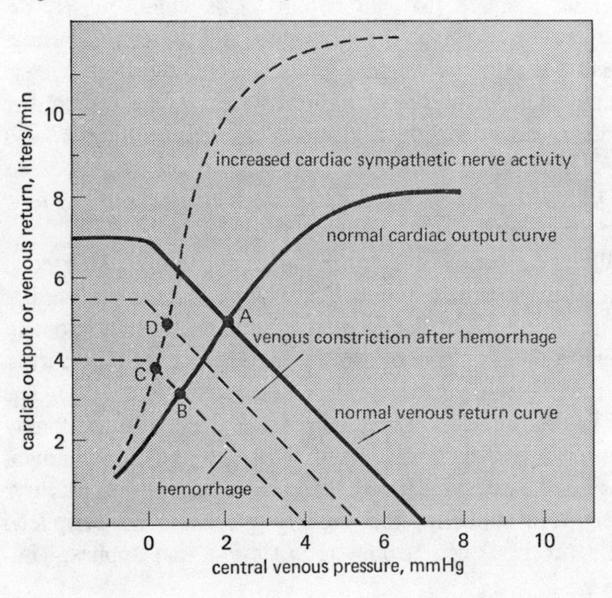

return falls below cardiac output at the central venous pressure of 2 mmHg. This is what leads to the fall in the central venous pool volume and pressure that causes the shift in operation from point A to point B. Note by comparing points A and B in Fig. 6-6 that blood loss itself lowers cardiac output and central venous pressure by shifting the venous return curve.

Subnormal cardiac output evokes a number of cardiovascular compensatory mechanisms in order to bring cardiac output back to more normal levels. One of these compensatory mechanisms is an increase in the activity of cardiac sympathetic nerves, and this shifts the heart's operation to a cardiac function curve that is higher than normal. The effect of increasing cardiac sympathetic activity is illustrated by a shift in cardiovascular operation from point B to point C. In itself, the increased cardiac sympathetic nerve activity increases cardiac output (from 3 to 4 liters/min) but causes a further decrease in central venous pressure. This drop in central venous pressure occurs because points B and C lie on the same venous return curve. *Cardiac* sympathetic nerves do not affect the venous return curve.[2]

An additional compensatory mechanism evoked by blood loss is increased activity of the sympathetic nerves leading to veins. Recall that this raises peripheral venous pressure and thus causes a rightward shift of the venous return curve. Therefore increased sympathetic activity to veins tends to shift the venous return curve, which was originally lowered by blood loss, back toward normal. As a consequence of the increased peripheral venous tone and the shift to a more normal venous return curve, the cardiovascular operation shifts from point C to point D in Fig. 6-6. Thus peripheral venous constriction increases cardiac output, and it does so by increasing central venous pressure and moving the heart's operation upward along a fixed cardiac function curve.

In summary, point D illustrates that normal cardiac output can be sustained in the face of blood loss by the combined effect of peripheral and cardiac adjustments. Hemorrhage is only one of an almost infinite variety of disturbances to the cardiovascular system. Plots such as those shown in Fig. 6-6 are very useful for understanding the many disturbances to the cardiovascular system and the ways in which they may be compensated.

Study questions: **31** to **33**

[2]Venous return is higher at point C than at point B, but the venous return curve has not shifted.

Chapter 7

Arterial Pressure and Its Regulation

OBJECTIVES

The student understands the measurement and determinants of arterial pressure:

1. Describes the auscultation technique of determining arterial systolic and diastolic pressures.
2. Identifies the physiological basis of the Korotkoff sounds.
3. Indicates the relationship between arterial pressure, cardiac output, and total peripheral resistance and predicts how arterial pressure will be altered when cardiac output and/or total peripheral resistance change.
4. Given arterial systolic and diastolic pressures, estimates mean arterial pressure.
5. Indicates the relationship between pulse pressure, stroke volume, and arterial compliance and predicts how pulse pressure will be changed by changes in stroke volume or arterial compliance.
6. Indicates the effects of altered cardiac contractility or total peripheral resistance on pulse pressure.

The student understands the mechanisms involved in the short-term regulation of arterial pressure:

7 Identifies the sensory receptors, afferent pathways, central integrating center, efferent pathways, and effector organs that participate in the baroreceptor reflex.
8 States the location of the arterial baroreceptors and describes their operation.
9 Describes the relationships between the vasomotor area and the cardioinhibitory area in the medullary cardiovascular center and how these areas are affected by altered baroreceptor input.
10 Diagrams the chain of events that are initiated by the baroreceptor reflex to compensate for a change in arterial pressure.
11 Describes the influence of central and peripheral chemoreceptors, intracranial pressure, low-pressure baroreceptors, and inputs from "higher centers" on the medullary cardiovascular center.
12 Describes and indicates the mechanisms involved in the cerebral ischemic response, the Cushing reflex, the alerting reaction, blushing, vasovagal syncopy, and the cardiovascular responses to pain.
13 Graphs the relationships between mean arterial pressure and sympathetic nerve activity that describe the overall operation of (1) the heart and peripheral vessels and (2) the baroreceptors plus the medullary cardiovascular center.
14 States what determines the normal mean arterial pressure and the normal level of sympathetic nerve activity.
15 Indicates how the relationship between sympathetic nerve activity and arterial pressure is shifted by a disturbance on the heart or vessels and how this alters the equilibrium within the baroreceptor reflex control system.
16 Indicates how the relationship between mean arterial pressure and sympathetic nerve activity is altered by nonarterial baroreceptor influences on the medullary cardiovascular center and how this shifts the equilibrium within the baroreceptor reflex control system.

The student understands the long-term regulation of arterial pressure:

17 Describes baroreceptor adaptation.
18 Describes the normal interrelationship between arterial pressure, urine output, and body fluid volume.
19 Indicates the mechanisms whereby altered arterial pressure alters glomerular filtration rate and renal tubular function to influence urine output.
20 Describes the relationship between fluid intake and the arterial pressure regulation of urine output.

Appropriate systemic arterial pressure is perhaps the single most important requirement for proper operation of the cardiovascular system. Without sufficient arterial pressure the brain and the heart do not receive adequate blood flow no matter what adjustments are made in their vascular resistance by local control mechanisms. On the other hand, unnecessary demands are placed on the heart by excessive arterial pressure. In this chapter we will discuss the determinants of arterial pressure and the elaborate mechanisms that have evolved for regulating this critical cardiovascular variable.

MEASUREMENT OF ARTERIAL PRESSURE

Recall that the systemic arterial pressure fluctuates with each heart cycle between a diastolic value (P_D) and a higher systolic value (P_S). Obtaining estimates of an individual's systolic and diastolic pressures is one of the most routine diagnostic techniques available to the physician. The basic principles of the *auscultation* technique used to measure blood pressure are described here with the aid of Fig. 7-1.

An inflatable cuff is wrapped around the upper arm and a device, such as a mercury manometer, is attached to monitor the pressure within the cuff. The cuff is initially inflated with air to a pressure that is well above normal systolic values (≃ 175 to 200 mmHg). This pressure is transmitted from the flexible cuff into the upper arm tissues, where it causes all blood vessels to collapse. No blood flows into (or out of) the forearm as long as the cuff pressure is higher than the systolic arterial pressure. After the initial inflation, air is allowed to gradually "bleed" from the cuff so that the pressure within it falls slowly and steadily through the range of arterial pressure fluctuations. The moment the

Figure 7-1 Blood pressure measurement by auscultation. Arrow A indicates systolic pressure. Arrow B indicates diastolic pressure.

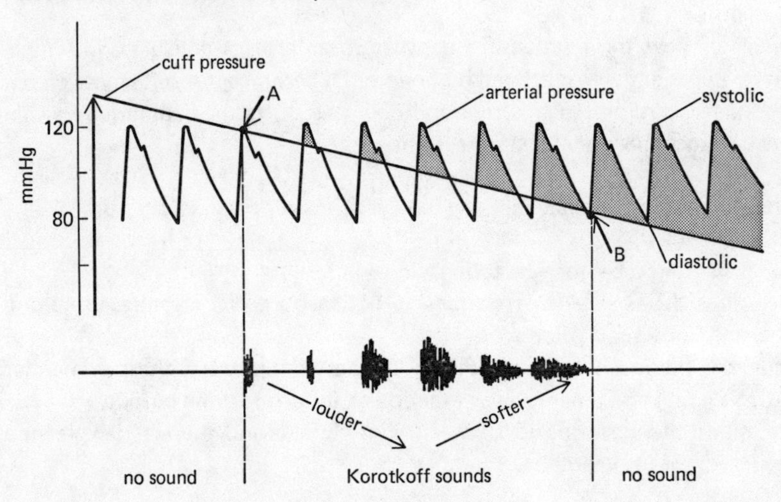

cuff pressure falls below the peak systolic arterial pressure, some blood is able to pass through the arteries beneath the cuff during the systolic phase of the cycle. This flow is intermittent and occurs only over a brief period of each heart cycle. Moreover, because it occurs through partially collapsed vessels beneath the cuff, the flow is turbulent rather than laminar. The intermittent periods of flow beneath the cuff produce tapping sounds, which can be detected with a stethoscope placed over the radial artery at the elbow. As indicated in Fig. 7-1, sounds of varying character, known collectively as *Korotkoff sounds*, are heard whenever the cuff pressure is between the systolic and diastolic aortic pressures.

Since there is no blood flow and thus no sound when cuff pressure is higher than systolic arterial pressure, *the highest cuff pressure at which tapping sounds are heard is taken as the systolic arterial pressure.* When the cuff pressure falls below the diastolic pressure, blood flows through the vessels beneath the cuff without periodic interruption and again no sound is detected over the radial artery. *The cuff pressure at which the sounds become muffled or disappear is taken as the diastolic arterial pressure.* The Korotkoff sounds are more distinct when the cuff pressure is near the systolic arterial pressure than when it is near the diastolic pressure. Thus consistency in determining diastolic pressure by auscultation requires concentration and experience.

DETERMINANTS OF ARTERIAL PRESSURE

Mean Arterial Pressure

Mean arterial pressure is a critically important cardiovascular variable because it indicates the average effective pressure that drives blood through the systemic organs. The mean arterial pressure that exists at any moment must equal that which causes the cardiac output (CO) to flow through the total peripheral resistance (TPR) of the systemic organs:

$$\bar{P}_A = \text{CO} \times \text{TPR}$$

This equation states one of the most fundamental relationships in cardiovascular physiology. Recognize that this is simply a rearrangement of the basic flow equation ($\Delta P = \dot{Q}R$) applied to the entire systemic circulation with the single assumption that centeral venous pressure is approximately zero so that $\Delta P = \bar{P}_A$. Note that mean arterial pressure is determined both by the heart (i.e., by cardiac output) and by the peripheral vasculature (by total peripheral resistance). All changes in mean arterial pressure result from changes in either cardiac output or total peripheral resistance. Furthermore, any intervention that alters mean arterial pressure must act by changing either cardiac output or total peripheral resistance.

Determining the true value of mean arterial pressure requires mathematically averaging the arterial pressure waveform over one or more complete heart cycles. Most often, however, we know only the systolic and diastolic pressures, yet wish to make some estimate of the mean arterial pressure. Mean arterial pressure necessarily falls between the systolic and diastolic pressures, but it is invariably found to be closer to diastolic pressure than systolic pressure because arterial pressure is near the diastolic value for a greater portion of the cardiac cycle. A common assumption is that mean arterial pressure ($\bar{P}_A$) is approximately equal to diastolic pressure (P_D) plus one-third of the difference between systolic and diastolic pressure ($P_S - P_D$):

$$\bar{P}_A = P_D + \tfrac{1}{3}(P_S - P_D)$$

Arterial Pulse Pressure

The *arterial pulse pressure* (P_p) is defined simply as systolic pressure minus diastolic pressure:

$$P_p = P_S - P_D$$

In previous chapters we discussed briefly how, as a consequence of the compliance of the arterial vessels, arterial pressure increases as arterial blood volume is expanded during cardiac ejection. The magnitude of the pressure increase (ΔP) caused by an increase in arterial volume depends on how large the volume change (ΔV) is and on how compliant (C_A) the arterial space is: $\Delta P = \Delta V/C_A$. If, as a first approximation, we neglect the fact that blood is leaving the arterial space *during* cardiac ejection, then the increase in arterial volume during each heartbeat is equal to the stroke volume (SV). Thus pulse pressure is approximately equal to stroke volume divided by arterial compliance:

$$P_p \simeq \frac{\text{SV}}{C_A}$$

Pulse pressure tends to increase with age because of decreased arterial compliance ("hardening of the arteries"). In the short term, however, the compliance of arterial vessels changes little, so that acute changes in pulse pressure are primarily the result of changes in stroke volume.

The equation above is based on the assumption that no blood leaves the aorta during systolic ejection. Obviously this is not strictly correct. Furthermore, close examination of Fig. 2-10 will reveal that peak systolic pressure is reached even before cardiac ejection is complete. It is therefore not surprising that several factors other than arterial compliance and stroke volume have minor influences on pulse pressure. For example, faster cardiac ejection caused by increased myocardial contractility tends to increase pulse pressure somewhat even if stroke volume remains constant. Changes in total peripheral resistance, however, have *little or no effect on pulse pressure*, since a change in TPR causes parallel changes in both systolic and diastolic pressure.

A common misconception in cardiovascular physiology is that the systolic pressure alone or the diastolic pressure alone indicates the status of a specific cardiovascular variable. For example, high diastolic pressure is often taken to indicate high total peripheral resistance. This is not necessarily so since high diastolic pressure can exist with normal (or even reduced) TPR if heart rate and cardiac output are high. Both systolic pressure and diastolic pressure are influenced by HR, SV, TPR, and C_A.[1]

SHORT-TERM INFLUENCES ON ARTERIAL PRESSURE

The arterial pressure is maintained at a level that ensures an adequate blood flow through the organs. Whenever the cardiovascular system is challenged, adjustments are made within the system that prevent large fluctuations in arterial pressure. In the short term (seconds), the necessary adjustments are made through changes in cardiac output and total peripheral resistance that are evoked by changes in the activity of the cardiovascular autonomic nerves. In the longer term (days), arterial pressure is controlled through changes in cardiac output produced by changes in blood volume. We will first consider nervous reflexes that control arterial pressure in the short term and how these cardiovascular reflexes may be modulated in special instances by neural influences from outside the major cardiovascular control area in the central nervous system.

Baroreceptor Reflex

The *baroreceptor reflex* is the most important mechanism providing short-term regulation of arterial pressure. Recall that the usual components of a reflex pathway include sensor receptors, afferent pathways, integrating centers in the central nervous system, efferent pathways, and effector organs. As shown in Fig. 7-2, the efferent pathways of the baroreceptor reflex are the cardiovascular sympathetic and cardiac parasympathetic nerves. The effector organs are the heart and peripheral blood vessels. We have not yet considered the sensory elements, the afferent pathways, or the integrating center in the brain stem that make the reflex complete.

[1] The equations presented in this chapter can be solved simultaneously to show that

$$P_S \simeq \text{SV} \times \text{HR} \times \text{TPR} + \frac{2}{3}\frac{\text{SV}}{C_A}$$

$$P_D \simeq \text{SV} \times \text{HR} \times \text{TPR} - \frac{1}{3}\frac{\text{SV}}{C_A}$$

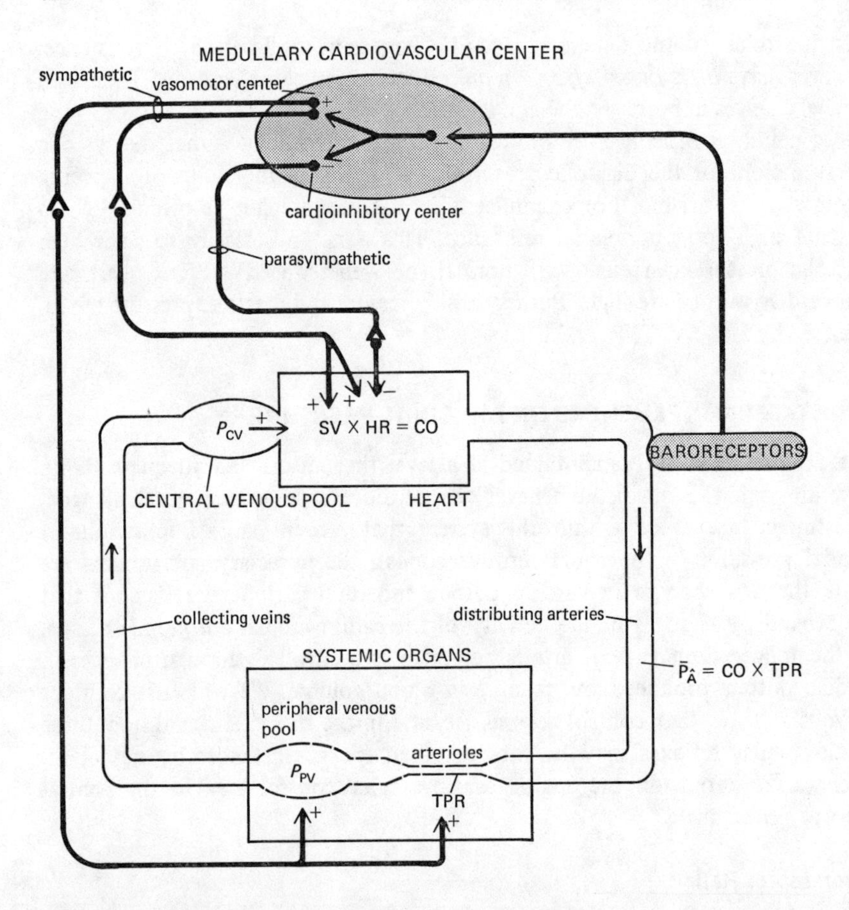

Figure 7-2 Components of the baroreceptor reflex pathway.

Baroreceptors Sensory receptors, called *baroreceptors*, are found in abundance in the walls of the aorta and carotid arteries. Major concentrations of these receptors are found near the arch of the aorta (the *aortic baroreceptors*) and at the bifurcation of the common carotid artery into the internal and external carotid arteries on either side of the neck (the *carotid sinus baroreceptors*). The receptors themselves are actually stretch receptors that detect changes in arterial pressure indirectly from the degree of stretch in the elastic arterial walls. The action potential generation rate of the baroreceptors increases when arterial pressure increases and decreases when arterial pressure falls.[2]

[2] Baroreceptor discharge rate can be enhanced by mechanical manipulation of the arterial walls. For example, the carotid sinus baroreceptor firing rate can be increased by massaging the neck over the carotid sinus area.

The relationship between arterial pressure and discharge rate varies considerably between individual baroreceptor sensory units. However, the collective discharge rate from many baroreceptors in the baroreceptor regions increases as mean arterial pressure increases in the range from 40 to about 180 mmHg. The overall baroreceptor discharge rate is most sensitive to changes in mean arterial pressure near the normal value of 100 mmHg. Baroreceptors do not discharge when mean arterial pressure is below 40 mmHg. The baroreceptor discharge rate is maximal when mean arterial pressure is about 180 mmHg or higher.

Baroreceptors respond to changes in pulse pressure as well as changes in mean arterial pressure. With a constant mean arterial pressure, baroreceptor firing rate is enhanced by an increase in pulse pressure.

If arterial pressure remains elevated over a period of several days for some reason, the baroreceptor firing rate will gradually return toward normal. Thus baroreceptors are said to *adapt* to long-term changes in arterial pressure. For this reason, the baroreceptor reflex cannot serve as a mechanism for the long-range regulation of arterial pressure.

Action potentials generated by the carotid sinus baroreceptors travel through the carotid sinus nerves (Hering's nerves), which join with the glossopharyngeal nerves (IXth cranial nerves) before entering the central nervous system. Afferent fibers from the aortic baroreceptors run to the central nervous system in the vagus nerves (Xth cranial nerves). (The vagus nerves contain both afferent and efferent fibers, including, for example, the parasympathetic efferent fibers to the heart.)

Medullary Cardiovascular Center The major area in the central nervous system that integrates and controls overall cardiovascular function is a region in the medulla oblongata of the brain stem called the *medullary cardiovascular center.* Anatomically, the cardiovascular center is rather diffuse and interconnected with other areas of the brain such as the adjacent respiratory control area. There are, however, two more or less distinct pools of nerve bodies within the cardiovascular center, as indicated in Fig. 7-2. The first area contains the nerve cell bodies of presynaptic cardiac and peripheral vascular sympathetic nerves and is sometimes called the *vasomotor center.* The second area contains the nerve bodies of the presynaptic cardiac parasympathetic fibers and is called the *cardioinhibitory center.*

The neurons in both the vasomotor area and the cardioinhibitory center area are spontaneously and tonically active. However, the levels of tonic neural activity in each area are modulated by outside influences on the cardiovascular control center.

As indicated in Fig. 7-2, the interconnections within the medullary cardiovascular center are such that there is normally a reciprocal relationship between the activities of the cardiovascular sympathetic and cardiac parasympathetic nerves. Almost invariably, an increase in activity of the cardiovascular sympathetic nerves is accompanied by a decrease in activity of the cardiac parasympathetic nerves and vice versa.

The major external influence on the medullary cardiovascular center comes from the arterial baroreceptors, as shown in Fig. 7-2. The neural interconnections are such that an increase in the baroreceptor discharge rate (increased arterial pressure) causes a decrease in the tonic activity of all cardiovascular sympathetic nerves and a simultaneous increase in the tonic activity of cardiac parasympathetic nerves. Conversely, decreased baroreceptor discharge causes increased sympathetic and decreased parasympathetic activity.

In this text we will consistently refer to any influence on the medullary cardiovascular center that decreases its sympathetic output and increases its parasympathetic output as a negative influence on the cardiovascular center. An increased arterial baroreceptor discharge rate then has a negative influence on the medullary cardiovascular center. Conversely, a decreased arterial baroreceptor discharge rate has a positive influence on the medullary cardiovascular center. All negative influences on the medullary cardiovascular center tend to produce a fall in arterial pressure, whereas all positive influences on the cardiovascular center tend to cause a rise in mean arterial pressure.

Operation of the Baroreceptor Reflex The baroreceptor reflex is a continuously operating control system that prevents large fluctuations in arterial pressure by making appropriate adjustments to any disturbance (stimulus). Figure 7-3 shows many events in the baroreceptor reflex pathway that occur in response to the stimulus of decreased mean arterial pressure. We have already discussed all of the events shown in Fig. 7-3, and each should be carefully examined (and reviewed if necessary) at this point because a great many of the interactions that are essential to understanding cardiovascular physiology are summarized in this figure.

Note in Fig. 7-3 that the response of the baroreceptor reflex to the stimulus of decreased mean arterial pressure is increased mean arterial pressure; i.e., the response tends to remove the stimulus. A stimulus of increased mean arterial pressure would elicit events exactly opposite to those shown in Fig. 7-3 and produce the response of decreased mean arterial pressure; again, the response tends to remove the stimulus. Thus the baroreceptor reflex is a *negative feedback mechanism* that operates automatically to resist changes in mean arterial pressure. The homeostatic benefits of the reflex action should be apparent.

The obvious consequence of the baroreceptor reflex is that the nervous control of the cardiovascular system operates rapidly to prevent changes in arterial pressure. Recall, however, that nervous control of vessels is more important in some organs like the kidney, the skin, and the splanchnic organs than in the brain and heart muscle. Thus the reflex response to a fall in arterial

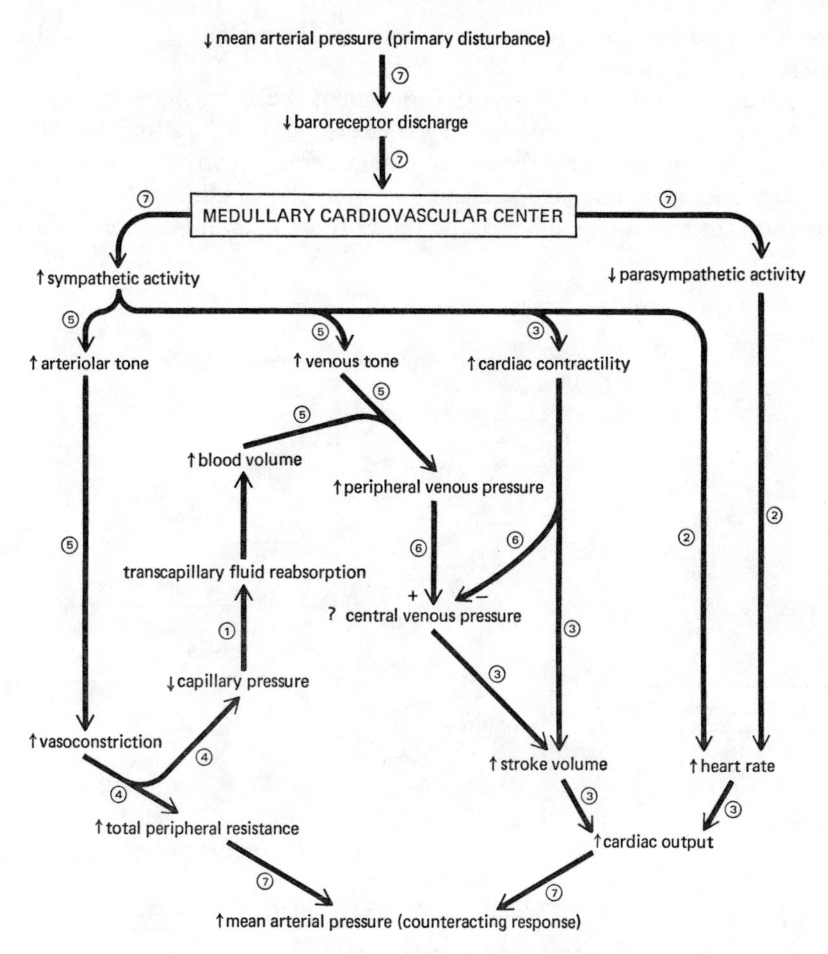

Figure 7-3 Immediate cardiovascular adjustments caused by a decrease in arterial blood pressure. Circled numbers indicate the chapter in which each interaction was discussed.

pressure may produce a significant increase in renal vascular resistance and a decrease in renal blood flow without changing the cerebral vascular resistance or blood flow. The peripheral vascular adjustments associated with the baroreceptor reflex take place primarily in the organs with strong sympathetic vascular control.

Other Cardiovascular Reflexes and Responses

Seemingly in spite of the baroreceptor reflex mechanism, large and rapid changes in mean arterial pressure occur in certain physiological and pathological situations. These reactions are caused by influences on the medullary cardiovascular center other than those from the arterial baroreceptors.

As indicated in Fig. 7-4, the medullary cardiovascular center receives input from many types of peripheral and central receptors as well as from so-called higher centers in the central nervous system through corticohypothalamic-medullary pathways. The precise manner in which these influences act on the neurons in the medullary cardiovascular center is not known. However, from a

Figure 7-4 Influences on the medullary cardiovascular center.

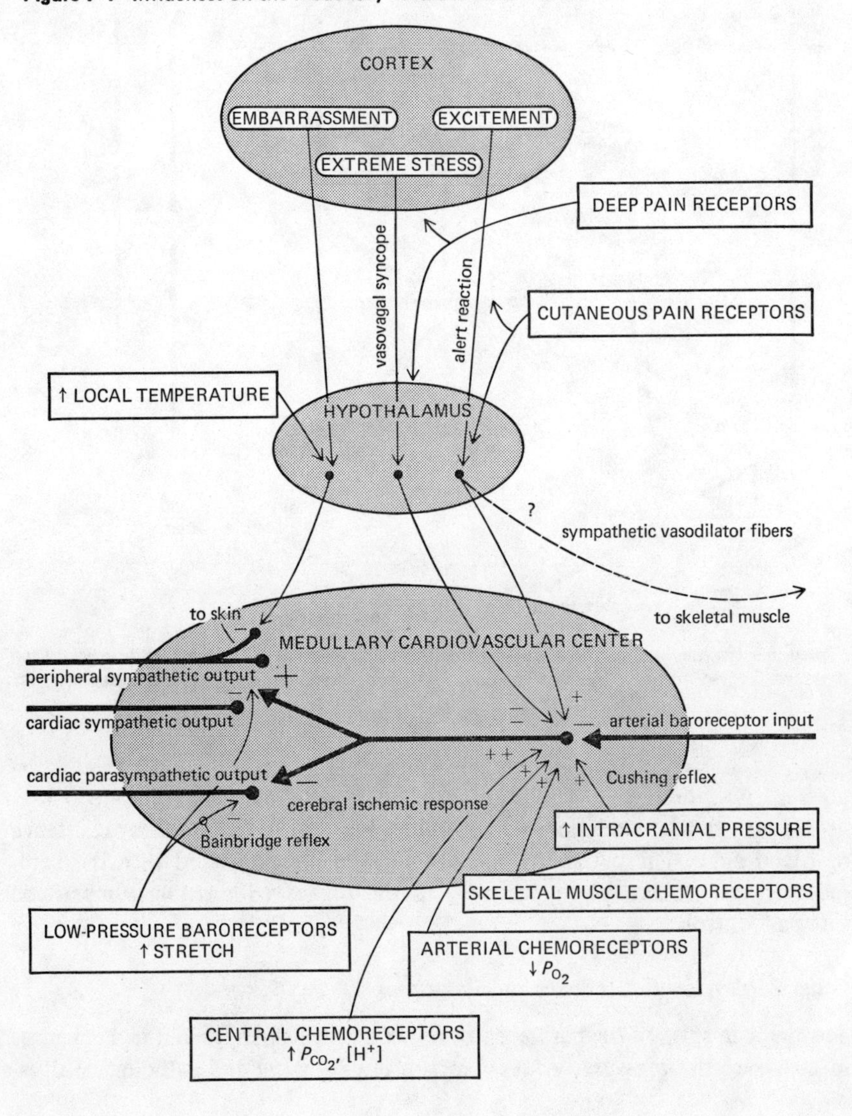

functional standpoint, we can understand the cardiovascular reactions they produce by assuming that they act within the medullary cardiovascular center at certain points along the basic pathway for integration of the arterial baroreceptor information as indicated in Fig. 7-4.

Certain nonbaroreceptor influences on the medullary cardiovascular center (those indicated by a + in Fig. 7-4) tend to cancel out some of the inhibitory influences on the medullary cardiovascular center from the arterial baroreceptors. This increases the level of sympathetic output from the cardiovascular center for any given level of input from the arterial baroreceptors. Thus a positive nonbaroreceptor influence on the medullary cardiovascular center causes the mean arterial pressure to be regulated to a higher than normal level by the baroreceptor reflex. Similarly, a negative nonbaroreceptor influence on the cardiovascular center would cause mean arterial pressure to be regulated to a lower than normal value.

Chemoreceptor Influences A marked rise in both respiratory rate and mean arterial pressure is caused by an increase in the P_{CO_2} or H^+ concentration of cerebrospinal fluid or by a decrease in the P_{O_2} of arterial blood. All these events are pathological because respiratory control mechanisms would normally preclude their occurrence.

An extremely strong autonomic reaction called the *cerebral ischemic response* is triggered by inadequate brain blood flow and can produce a more intense sympathetic vasoconstriction and cardiac stimulation than is elicited by any other influence on the cardiovascular control center. Presumably the cerebral ischemic response is mediated by chemoreceptors located within the central nervous system as indicated in Fig. 7-4. If cerebral blood flow is severely inadequate for several minutes, the cerebral ischemic response wanes and is replaced by marked loss of sympathetic activity. Presumably this situation results when function of the nerve cells in the cardiovascular center become directly depressed by the unfavorable chemical conditions in the cerebrospinal fluid.

Whenever intracranial pressure is increased—for example, by tumor growth within the rigid cranium—there is a parallel rise in arterial pressure. This is called the *Cushing reflex.* It can cause mean arterial pressures of more than 200 mmHg in severe cases of intracranial pressure elevation. The obvious benefit of the Cushing reflex is that it prevents collapse of cranial vessels and thus preserves adequate brain blood flow even in spite of large increases in intracranial pressure. The mechanisms responsible for the Cushing reflex are not known but could involve the central chemoreceptors.

There is evidence suggesting that some type of chemoreceptor exists in skeletal muscle that has a positive influence on blood pressure. This possibility is indicated in Fig. 7-4 by dashed lines. It is suggested that these receptors play a role in the rise in mean blood pressure that accompanies strenuous exercise.

Influences from Low-Pressure Receptors Stretch receptors, similar to those in systemic arteries, are found in the atria and the pulmonary arteries. These are called *low-pressure receptors* to distinguish them from the systemic arterial baroreceptors. The role of low-pressure receptors in the operation of the cardiovascular system is not as well established as that of high-pressure receptors. Distension of the low-pressure receptors causes a decrease in sympathetic tone to peripheral vessels just as does stretch of systemic arterial baroreceptors. Rapid atrial stretch, however, produces a marked *increase* in heart rate due in part to a reflex mechanism known as the *Bainbridge reflex.*

Many investigators believe that low-pressure receptors are more important to fluid volume reflexes involving the kidney than to short-term blood pressure regulation.

Cardiovascular Responses Associated with Emotion Cardiovascular responses are frequently associated with certain states of emotion. Presumably these responses originate in the cerebral cortex and reach the medullary cardiovascular center through the corticohypothalamic pathways, as shown in Fig. 7-4. The least complicated of these responses is the *blushing* that is often detectable in individuals with lightly pigmented skin during states of embarrassment. The blushing response involves a loss of sympathetic vasoconstrictor activity *only* to cutaneous vessels, and this produces the blush by allowing engorgement of the cutaneous venous sinus. Embarrassment presumably causes this discrete response by inhibiting the tonic activity of certain sympathetic vasoconstrictor nerves by an action somewhat like that shown in Fig. 7-4.

Excitement or a sense of danger often elicits a complex behavioral pattern called the *alerting reaction* (sometimes called the "fight or flight" response). The alerting reaction involves a host of responses such as pupillary dilation and increased skeletal muscle tenseness, which are generally appropriate preparations for some form of intense physical activity. The cardiovascular component of the alerting reaction is an increase in blood pressure caused by a general increase in cardiovascular sympathetic nervous activity and a decrease in cardiac parasympathetic activity. These cardiovascular effects are caused by a positive influence on the medullary cardiovascular control center from the corticohypothalamic pathways. A dashed line in Fig. 7-4 is used to signify the possible activation of sympathetic cholinergic vasodilator fibers to skeletal muscle during the alerting reaction. (Recall, however, that it is not certain that sympathetic vasodilator fibers exist in humans.)

Some individuals respond to situations of extreme stress by fainting, and the overall process is sometimes called *vasovagal syncope.* The loss of consciousness is due to decreased cerebral blood flow, which is itself produced by a sudden dramatic loss of arterial blood pressure. Figure 7-4 shows how the corticohypothalamic pathways might influence the cardiovascular center to produce the sudden loss of sympathetic tone and increase in parasympathetic tone that occur with vasovagal syncope. Fortunately, unconsciousness appears to quickly remove this strong negative influence on the cardiovascular center.

The extent to which cardiovascular variables, in particular blood pressure, are normally affected by emotional state is currently a topic of extreme interest and considerable research. As yet the answer is unclear. However, the therapeutic value of being able, for example, to learn to control one's blood pressure would be incalculable.

Cardiovascular Responses to Pain As indicated in Fig. 7-4, pain can have either a positive or a negative influence on the cardiovascular center and thus a positive or a negative influence on arterial pressure. Generally, superficial or cutaneous pain causes a rise in blood pressure in a manner similar to that associated with the alerting response and perhaps over many of the same pathways. Deep pain from receptors in the viscera or joints, however, often causes a cardiovascular response similar to that which accompanies vasovagal syncope, i.e., a serious decrease in blood pressure. This response may contribute to the state of shock that often accompanies crushing injuries and/or joint displacement.

Temperature Regulation Reflexes Certain special cardiovascular reflexes that involve the control of skin blood flow have evolved as part of the body temperature regulation mechanisms. Temperature regulation responses are controlled primarily by the hypothalamus, and as shown in Fig. 7-4 the hypothalamus can operate through the cardiovascular center to control the sympathetic activity to cutaneous vessels and thus skin blood flow. The cardiovascular center is extremely responsive to changes in hypothalamic temperature. Measurable changes in cutaneous blood flow result from changes in hypothalamic temperature of tenths of a degree Celsius.

Cutaneous vessels are influenced by reflexes involved in both arterial pressure regulation and temperature regulation. When the appropriate cutaneous vascular responses for temperature regulation and pressure regulation are contradictory, as they are, for example, during strenuous exercise in a hot environment, then the temperature-regulating influences on cutaneous blood vessels prevail.

Equilibrium in the Baroreceptor Control Systems

We can conceptualize, as shown in Fig. 7-5, that the complete baroreceptor reflex pathway is a control system made up of two distinct portions: (1) the *effector portion*, including the heart and peripheral blood vessels, and (2) the *neural control portion*, which includes the arterial baroreceptors, their afferent nerve fibers, the medullary cardiovascular center, and the efferent sympathetic and parasympathetic fibers. Mean arterial pressure is the output of the effector

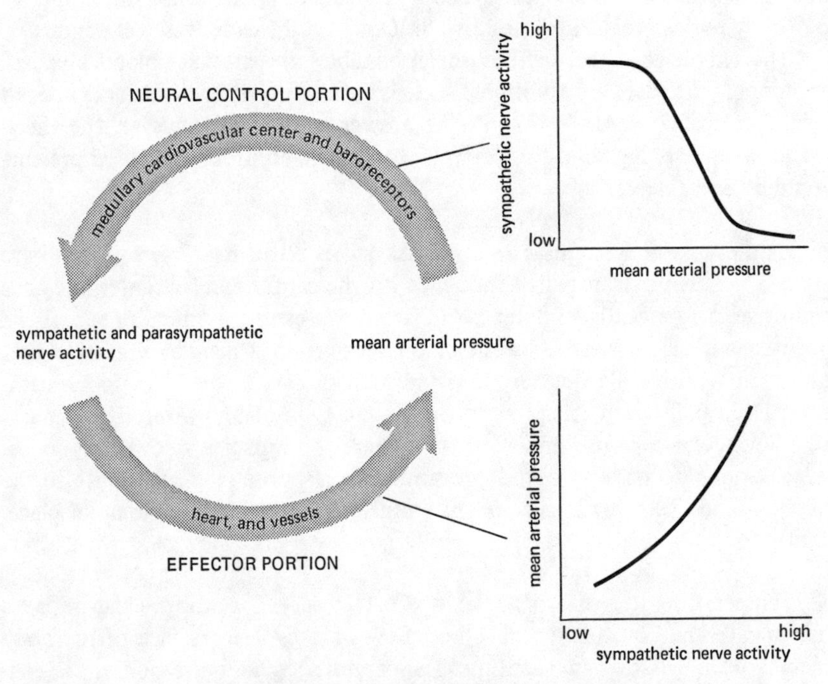

Figure 7-5 Neural control and effector portions of the baroreceptor control system.

portion *and simultaneously* the input to the neural control portion. Similarly, the activity of the sympathetic (and parasympathetic)[3] cardiovascular nerves is the output of the neural control portion of the baroreceptor control system and, at the same time, the input to the effector portion.

In Chaps. 3, 5, and 7 we discussed a host of reasons why mean arterial pressure *increases* when the heart and peripheral vessels receive *increased* sympathetic nerve activity. All this information can be summarized by the curve shown in the lower graph of Fig. 7-5, which describes how the effector portion of the baroreceptor control system functions.

We have also discussed in this chapter how *increased* mean arterial pressure acts through the baroreceptors to inhibit the medullary cardiovascular center and to *decrease* the sympathetic activity. Thus, for the neural control portion of the baroreceptor control system *alone*, increased arterial pressure produces decreased sympathetic nerve activity, as summarized by the upper graph in Fig. 7-5.

When the baroreceptor control system is intact, its effector portion and

[3] For convenience, we will omit continual reference to parasympathetic nerve activity in the following discussion. Throughout, however, an indicated change in sympathetic nerve activity should also be taken to imply a reciprocal change in the activity of the cardiac parasympathetic nerves.

neural control portion must interact until they come into equilibrium. This will occur at some combination of mean arterial pressure and sympathetic activity that is compatible with the way both portions of the system operate. To determine what the equilibrium values of mean arterial pressure and sympathetic nerve activity will be, we can plot the individual function curves for the effector and neural control portions of the baroreceptor control system together on the same graph as in Fig. 7-6A. The only point in Fig. 7-6A that is compatible with the way both the effector portion and the neural control portion operate

Figure 7-6 Operation of the baroreceptor control system. A. Normal equilibrium. B. Equilibrium shift with disturbance on the effector portion.

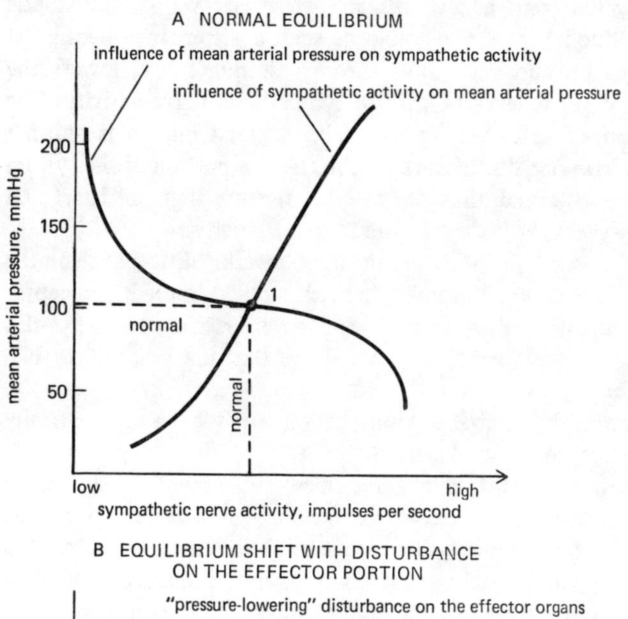

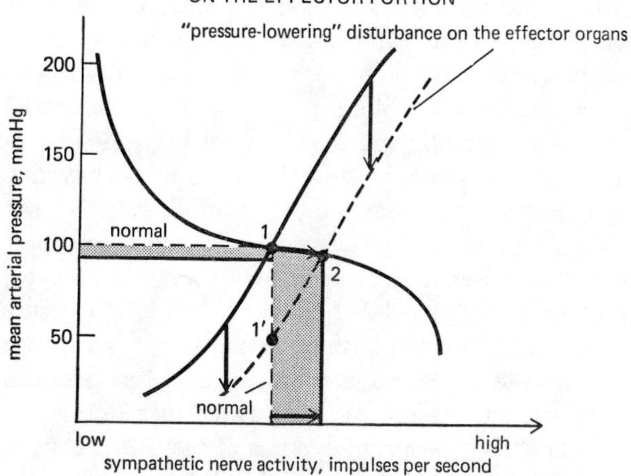

is the point where their operating curves intersect. Therefore, the normal mean arterial pressure and the normal level of sympathetic activity are determined by the equilibrium point of operation established within the baroreceptor control system.

Whenever there is any outside disturbance on the cardiovascular system, the equilibrium within the baroreceptor control system shifts away from normal. This happens because *all* cardiovascular disturbances cause a shift in one or the other of the two curves in Fig. 7-6A. For example, Fig. 7-6B shows how the equilibrium for the baroreceptor control system is shifted away from normal by a cardiovascular disturbance that lowers the operating curve of the effector portion. The disturbance in this case could be anything that reduces the arterial pressure produced by the heart and vessels at each given level of sympathetic input they receive. Blood loss, for example, is such a disturbance because it lowers central venous pressure and thus, through Starling's law, lowers the cardiac output at any given level of cardiac sympathetic nerve activity. The metabolic vasodilation of arterioles in exercising skeletal muscle is another example of a pressure-lowering disturbance on the effector portion of the system because it lowers the TPR and thus the arterial pressure that the heart and vessels produce at any given level of sympathetic nerve activity.

As shown by point 2 in Fig. 7-6B, any pressure-lowering disturbance on the heart or vessels causes a new equilibrium to be reached within the baroreceptor control system at a slightly lower than normal mean arterial pressure and a higher than normal sympathetic activity level. Note that the point 1′ in Fig. 7-6B indicates how far the mean arterial pressure would have fallen as a consequence of the disturbance had not the sympathetic activity been automatically increased by the baroreceptor control system.

As indicated previously in this chapter, many disturbances act on the neural portion of the baroreceptor control system rather than directly on the heart or vessels. These disturbances shift the equilibrium within the cardiovascular system away from normal because they alter the operating curve of the neural control portion of the system. For example, some nonbaroreceptor influences on the medullary cardiovascular center (those indicated by a + in Fig. 7-4) act to cancel out some of the inhibitory influence from the arterial baroreceptors. This increases the level of sympathetic output from the cardiovascular center for any given level of arterial pressure; i.e., it shifts the operating curve for the neural control portion of the baroreceptor control system to the right, as shown in Fig. 7-7A. Thus when there is a positive nonbaroreceptor influence on the cardiovascular center, the components of the baroreceptor control system reach equilibrium at a higher than normal arterial pressure and a higher than normal sympathetic activity. In a similar manner, a negative nonbaroreceptor influence on the cardiovascular center (as with vasovagal syncopy) would cause mean arterial pressure to be regulated to a lower than normal value because it would shift the operating curve of the neural control portion of the baroreceptor control system to the left in Fig. 7-7A.

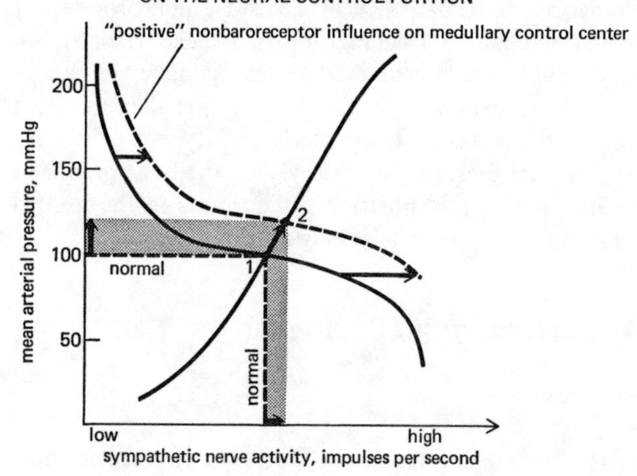

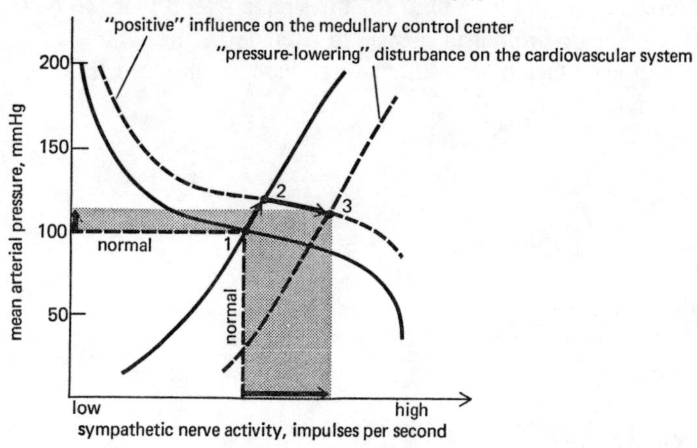

Figure 7-7 Effect of neural influences on the baroreceptor control system A. Equilibrium shift with disturbance on the neural control portion. B. Equilibrium shift with disturbances on both neural control and effector portions.

Figure 7-7B shows how the equilibrium values will be changed when a pressure-lowering disturbance on the heart or vessels is added to a positive nonbaroreceptor influence on the medullary cardiovascular center. The change in equilibrium from point 1 to point 2 in Fig. 7-7B shows how the baroreceptor control system responds to the positive nonbaroreceptor influence on the cardiovascular center alone. Superimposing a pressure-lowering disturbance on the heart or vessels shifts the equilibrium from point 2 to point 3. Note that, although the response to the pressure-lowering disturbance in Fig. 7-7B (point

2 to point 3) starts from a higher than normal arterial pressure, it is essentially identical to that which occurs in the absence of a positive nonbaroreceptor influence on the cardiovascular center (see Fig. 7-6B). Nonbaroreceptor influences on the medullary cardiovascular center do not change how the cardiovascular system responds to disturbances on it, they simply change the starting point. The fact that higher than normal sympathetic activity can exist with higher than normal mean arterial pressure does not violate the basic principles of the baroreceptor reflex. In fact, this situation often occurs in the normal operation of the cardiovascular system, as will be illustrated in Chap. 8.

LONG-TERM REGULATION OF ARTERIAL PRESSURE

Fluid Balance and Arterial Pressure

We have already considered several key factors in the long-term regulation of arterial blood pressure. First is the fact that the baroreceptor reflex, however well it counteracts temporary disturbances in arterial pressure, cannot effectively regulate arterial pressure in the long term for the simple reason that the baroreceptor firing rate adapts to prolonged changes in arterial pressure.

The second pertinent fact is that circulating blood volume can influence arterial pressure because

↑ Blood volume
↓
↑ Peripheral venous pressure
↓
Right shift of venous return curve
↓
↑ Central venous pressure
↓
↑ Cardiac output
↓
↑ Arterial pressure

A fact we have yet to consider is that arterial pressure has a profound influence on urine output rate and thus affects total body fluid volume. Since blood volume is one of the components of the total body fluid, blood volume alterations accompany changes in total body fluid volume. The mechanisms are such that an *increase in arterial pressure* causes an increase in urine output rate and thus a *decrease in blood volume.*

The negative influence of arterial pressure on blood volume forms one component of the *fluid balance regulatory feedback mechanism* regulating arterial pressure, as shown on the right side of Fig. 7-8. Just as with the baroreceptor reflex, shown on the left side of Fig. 7-8, the fluid balance feedback loop acts to correct disturbances in arterial pressure. For example:

↑ Arterial pressure (disturbance)
↓
↑ Urine output rate
↓
↓ Fluid volume
↓
↓ Blood volume
↓
↓ Cardiac output
↓
↓ Arterial pressure (compensation)

Both the baroreceptor reflex and the fluid balance mechanism are negative feedback loops that regulate arterial pressure. Whereas the baroreceptor reflex is very quick to counteract disturbances in arterial pressure, hours or even days may be required before a change in urine output rate produces a significant accumulation or loss of total body fluid volume. Whatever the fluid balance mechanism lacks in speed, however, it more than makes up for in persistence. As long as there is *any* inequality between the fluid intake rate and the urine

Figure 7-8 Mechanisms of short- and long-term regulation of arterial pressure.

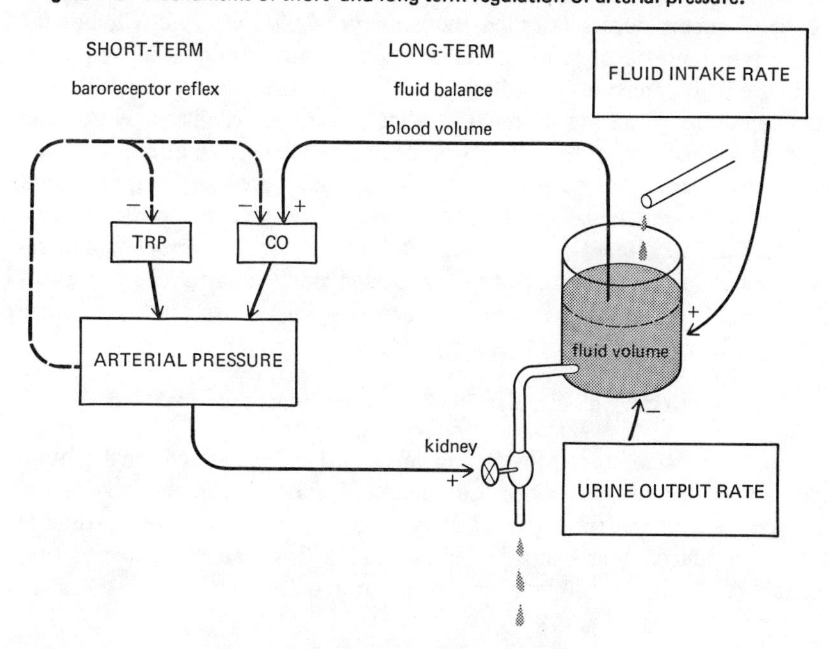

output rate, fluid volume is changing and the fluid balance mechanism has not completed its adjustment of arterial pressure. The fluid balance mechanism is in equilibrium only when the urine output rate exactly equals the fluid intake rate. *In the long term, the arterial pressure can only be that which makes the urine output rate equal to the fluid intake rate.*

The baroreceptor reflex is, of course, essential for counteracting rapid changes in arterial pressure. The fluid balance mechanism, however, determines the long-term level of arterial pressure because it slowly overwhelms all other influences. Through adaptation, the baroreceptor mechanism adjusts itself so that it operates to prevent acute changes in blood pressure from the prevailing long-term level as determined through fluid balance.

Effect of Arterial Pressure on Urine Output Rate

A key element in the fluid balance mechanism of arterial pressure regulation is the effect that arterial pressure has on the renal urine production rate. The mechanisms responsible for this will be only briefly described here with emphasis on their cardiovascular implications.

As indicated in Chap. 1, the kidneys play a major role in homeostasis by regulating the electrolyte composition of the plasma and thus the entire internal environment. One of the major plasma electrolytes regulated by the kidneys is the sodium ion. To regulate the electrolyte composition, a large fraction of the plasma fluid that flows into the kidneys is filtered across the *glomerular capillaries* so that it enters the *renal tubules.* The fluid that passes from the blood into the renal tubules is called the *glomerular filtrate* and the rate at which this process occurs is called the *glomerular filtration rate.* Glomerular filtration is a transcapillary fluid movement whose rate is influenced by hydrostatic and oncotic pressures as indicated in Chap. 1. The primary cause of continual glomerular filtration is the fact that glomerular capillary hydrostatic pressure is normally very high ($\simeq$70 mmHg). The glomerular filtration rate is decreased by factors that decrease glomerular capillary pressure, e.g., decreased arterial blood pressure or vasoconstriction of preglomerular renal arterioles.

Once fluid is filtered into the renal tubules, it (1) is *reabsorbed* and reenters the cardiovascular system, or (2) is passed along renal tubules and eventually *excreted* as urine. Thus urine production is the net result of glomerular filtration and renal tubular fluid reabsorption:

$$\text{Urine output rate} = \text{glomerular filtration rate} - \text{renal fluid reabsorption rate}$$

Actually, most of the reabsorption of fluid that has entered renal tubules as glomerular filtrate occurs because sodium is actively pumped out of the tubules by cells in the tubular wall. When sodium leaves the tubules, osmotic forces are produced that cause water to leave with it. Thus any factor that promotes renal tubular sodium reabsorption (sodium retention) tends to increase the renal fluid reabsorption rate and consequently decrease the urine output rate. The blood concentration of the hormone *aldosterone*, which is produced by the adrenal glands, is the primary regulator of the rate of sodium reabsorption by renal tubular cells. Adrenal release of aldosterone is, in turn, regulated largely by the circulating level of another hormone, *angiotensin II*, which forms in the plasma under the regulation of an enzyme, *renin*, that is produced by the kidneys.

The rate of renin release by the kidneys appears to be influenced by several factors. An increase in the activity of renal sympathetic nerves causes the release of renin through a beta-adrenergic mechanism. Also, renin release is triggered by factors associated with a lowered glomerular filtration rate. The important fact to keep in mind, from a cardiovascular standpoint, is that anything that causes renin release causes a decrease in urine output rate because increased renin causes increased sodium (and therefore fluid) reabsorption from renal tubules.

Urine output rate is also influenced by a hormone released from the poster-

Figure 7-9 Mechanisms by which arterial pressure influence urine output rate.

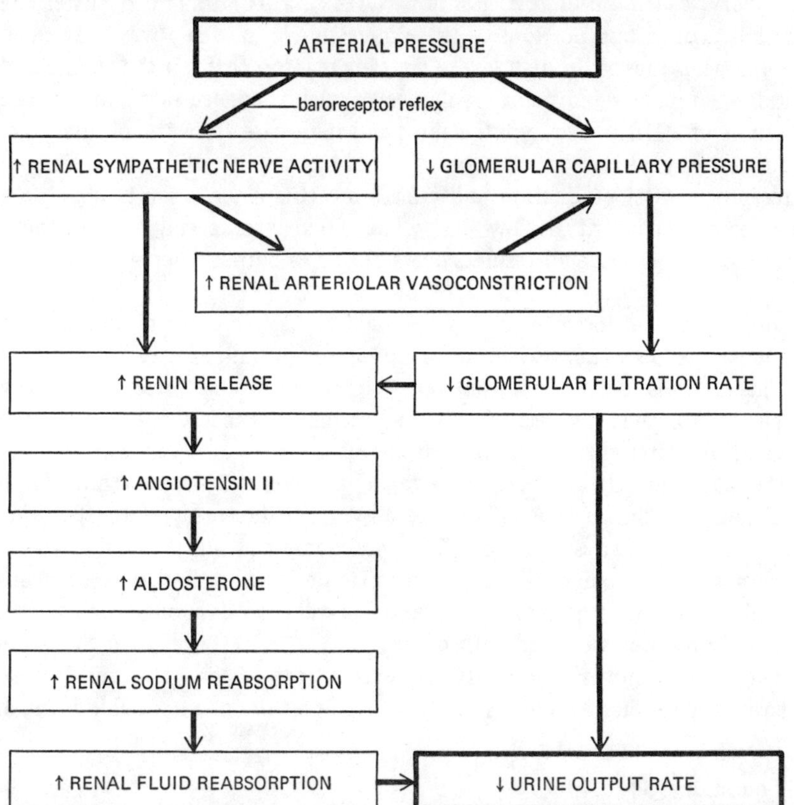

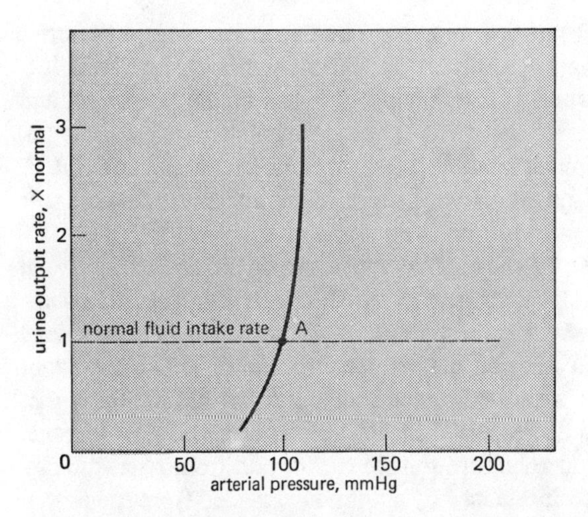

Figure 7-10 Effect of arterial pressure on urine output rate in a normal person.

ior pituitary, called antidiuretic hormone (ADH). This hormone regulates the permeability of certain portions of the kidney tubule in such a way that when the hormone is present in high levels, water is reabsorbed from the tubule and the kidney produces only small volumes of highly concentrated urine. The production of ADH in the hypothalamus and its release from the pituitary are influenced by many factors, one of which is the activity of the low-pressure baroreceptors. ADH production and release are stimulated when the low-pressure baroreceptor activity is low–i.e., when blood volume and central venous volume are low. The overall result is decreased urine output when blood volume is low.

Some of the major mechanisms that lead to decreased urine output rate are summarized in Fig. 7-9. The most important information to be obtained from this figure is that urine output rate is linked to arterial pressure by many synergistic pathways. Because of this, modest changes in arterial pressure are associated with large changes in urine output rate.

The observed relation between arterial pressure and urine output for a normal person is shown in Fig. 7-10. Recall that, in the steady state, the urine output rate must always equal the fluid intake rate and that the fluid balance mechanisms will automatically adjust arterial pressure until this is so. Thus a normal person with a normal fluid intake rate will have, as a long-term average, the arterial pressure associated with point A in Fig. 7-10. Because of the steepness of the curve shown in Fig. 7-10, even rather marked changes in fluid intake rate have rather minor influences on the arterial pressure of a normal individual.

Study questions: **34** to **43**

Chapter 8

Cardiovascular Responses to Homeostatic Disturbances

OBJECTIVES

The student understands the mechanisms involved in the cardiovascular responses to homeostatic disturbances on the intact cardiovascular system and can predict the resulting alterations in all important cardiovascular variables:

1 States how gravity influences arterial, venous, and capillary pressures at any height above or below the heart in a standing individual.
2 Describes and explains the changes in central venous pressure and the changes in transcapillary fluid balance and venous volume in the lower extremities caused by standing upright.
3 Describes the operation of the "skeletal muscle pump" and explains how it simultaneously promotes venous return and decreases capillary hydrostatic pressure in the muscle vascular beds.
4 States the changes in central venous pressure that are associated with respiratory movements.
5 Describes how the "respiratory pump" promotes venous return.

6 Defines "decompensatory mechanism" and gives several examples of decompensatory processes that can occur in the cardiovascular system.

7 Defines shock, heart failure, and hypertension.

8 Describes the cardiovascular alterations accompanying *any* homeostatic disturbance by:

- **a** Identifying the primary disturbances that the situation places on the cardiovascular system.
- **b** Listing how the primary disturbances change the influence on the medullary cardiovascular center from (1) arterial baroreceptors and (2) other sources.
- **c** Stating what changes will occur in sympathetic and parasympathetic nerve activity as a result of the altered influences on the medullary cardiovascular center.
- **d** Indicating what immediate reflex changes will occur in heart rate, cardiac contractility, stroke volume, arteriolar tone, venous tone, peripheral venous pressure, central venous pressure, total peripheral resistance, resistance in any major organ, and blood flow through any major organ.
- **e** Predicting what the net effect on mean arterial pressure of the primary and reflex influences on the cardiovascular variables listed in objective 8d will be.
- **f** Stating whether mean arterial pressure and sympathetic nerve activity will settle above or below their normal values.
- **g** Predicting whether and stating how cutaneous blood flow will be altered by temperature regulation reflexes.
- **h** Indicating whether and how transcapillary fluid movements will be involved in the overall cardiovascular response.
- **i** Indicating whether, why, how, and with what time course renal adjustments of fluid balance will participate in the response.
- **j** Predicting how each of the variables listed in objective 8d will be influenced by long-term adjustments in blood volume.
- **k** Indicating whether and how decompensatory mechanisms affect the cardiovascular response.

In this chapter we will see how the basic principles of cardiovascular physiology, which have been discussed, apply to the intact cardiovascular system. A variety of situations that tend to disturb homeostasis will be presented. The key to understanding the cardiovascular adjustments in each situation is to recall that the baroreceptor reflex always acts to blunt changes in arterial pressure. The overall result is that *adequate blood flow to the brain and the heart muscle is maintained in any circumstance.*

The cardiovascular alterations in each of the following examples are produced by the combined effects of (1) the primary, direct influences of the disturbance on the cardiovascular variables and (2) the reflex adjustments that are triggered by the primary disturbances. The general pattern of reflex adjustment is similar in all situations. Rather than trying to memorize the cardiovascular alterations that accompany each situation, the student should strive to understand each response in terms of the primary disturbances and reflex reactions involved.

An extensive list of study questions is supplied for this chapter. These questions are intended to reinforce the student's understanding of complex cardiovascular responses and provide a review of basic cardiovascular principles.

CHANGES IN BODY POSITION

Because gravity has an effect on pressures within the cardiovascular system, significant cardiovascular readjustments accompany changes in body position. In the preceding chapters, the influence of gravity was ignored and pressure differences between various points in the systemic circulation were related only to flow and vascular resistance ($\Delta P = \dot{Q}R$). As shown in Fig. 8-1, this is approximately true only for a recumbent individual. In a standing individual, additional cardiovascular pressure differences exist between the heart and regions that are not at heart level. The most extreme situation occurs in the lower legs and feet of a standing individual. As indicated in Fig. 8-1B, all intravascular pressures in the feet may be increased by 90 mmHg simply from the weight of the blood in the arteries and veins leading to and from the feet. Note by comparing Fig. 8-1A and B that standing upright does not in itself change the flow through the lower extremities, since gravity has the same effect on arterial and venous pressures and thus does not change the *arteriovenous pressure difference* at any one height level. There are, however, two major direct effects of the increased pressure in the lower extremities that are shown in Fig. 8-1B: (1) the absolute increase in venous pressure distends peripheral veins and greatly increases peripheral venous volume and (2) the absolute increase in capillary hydrostatic pressure causes a tremendously high transcapillary filtration rate.

For reasons to be described, a reflex activation of sympathetic nerves accompanies the transition from a recumbent to an upright position. However, Fig. 8-1C shows how vasoconstriction from sympathetic activation is only marginally effective in ameliorating the adverse effects of gravity on the lower extremities. Arteriolar constriction can cause a greater pressure drop across arterioles, but this has only a limited effect on capillary pressure because venous pressure remains extremely high. Filtration will continue at a very high rate. In fact, the normal cardiovascular reflex mechanisms alone are incapable of dealing with upright posture without the aid of the "skeletal muscle pump." A person who remained upright without intermittent contraction of the skeletal muscles in the legs would lose consciousness in 10 to 20 min because of the decreased brain blood flow that would stem from diminished central blood volume, stroke volume, cardiac output, and arterial pressure.

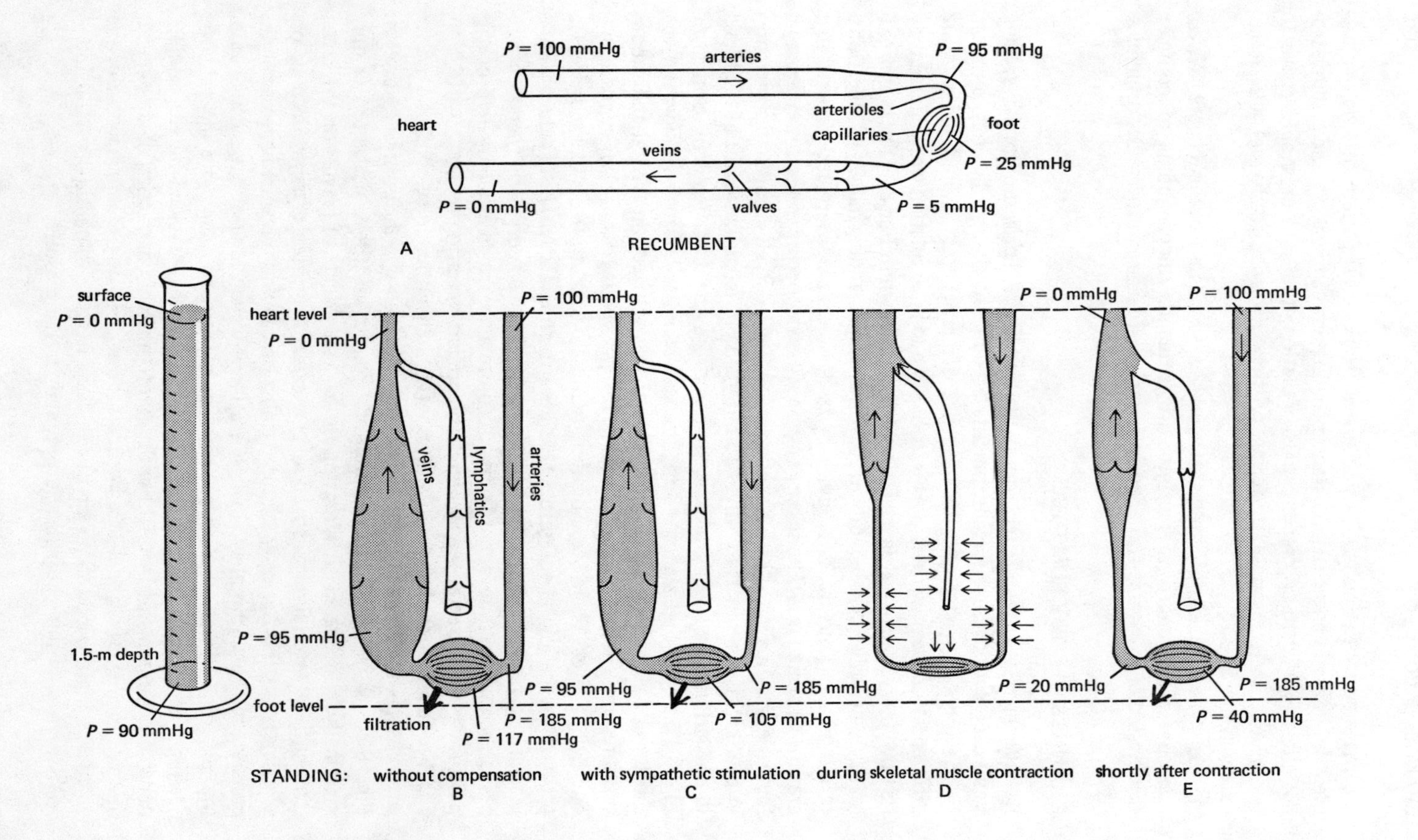

Figure 8-1 Effect of gravity on vascular pressures (A and B) with compensatory influences of sympathetic stimulation (C) and the skeletal muscle pump (D and E).

The effectiveness of the skeletal muscle pump in counteracting venous blood pooling and edema formation in the lower extremities during standing is illustrated in Fig. 8-1D and E. The compression of vessels during skeletal muscle contraction expels both venous blood and lymphatic fluid from the lower extremities (Fig. 8-1D). Immediately after a skeletal muscle contraction, both veins and lymphatic vessels are relatively empty because their one-way valves prevent the back flow of previously expelled fluid (Fig. 8-1E). Most importantly, the weight of the venous and lymphatic fluid columns is temporarily supported by the closed one-way valve leaflets. Consequently, venous pressure is drastically lowered immediately after skeletal muscle contraction and rises only gradually as veins refill with blood from the capillaries. Thus capillary pressure and transcapillary fluid filtration rate are dramatically reduced for some period after a skeletal muscle contraction. Periodic skeletal muscle contractions can keep the average value of venous pressure at levels that are only moderately above normal. This, in combination with an increased pressure drop across vasoconstricted arterioles, prevents capillary pressures from rising to intolerable levels in the lower extremities. Some transcapillary fluid filtration is still present, but the increased lymphatic flow resulting from the skeletal muscle pump is normally sufficient to prevent severe edema formation in the feet.

The actions of the skeletal muscle pump, however beneficial, do not completely prevent a rise in the average venous pressure and blood pooling in the lower extremities on standing. Thus, assuming an upright position upsets the cardiovascular system and elicits reflex cardiovascular adjustments, as shown in Fig. 8-2.

As with all cardiovascular responses, the key to understanding the alterations associated with standing is to distinguish the primary disturbances from the compensatory responses. As shown in Fig. 8-2, the immediate consequence of standing is an increase in both arterial and venous pressure in the lower extremities. By the chain of events shown, the primary disturbances produce a net "positive influence" on the cardiovascular center by lessening the normal inhibitory influences from both the arterial and the low-pressure baroreceptors.

The result of a net positive influence on the cardiovascular center will be reflex adjustments appropriate to increase blood pressure, i.e., decreased cardiac parasympathetic nerve activity and increased activity of the cardiovascular sympathetic nerves. Heart rate and cardiac contractility will increase, as will arteriolar and venous constriction in most systemic organs (brain and heart excepted).[1]

Heart rate and total peripheral resistance are higher when an individual stands than when the individual is lying down. Note that these particular cardio-

[1] Each of these steps should be familiar to you at this point. Please examine each step carefully and review as necessary.

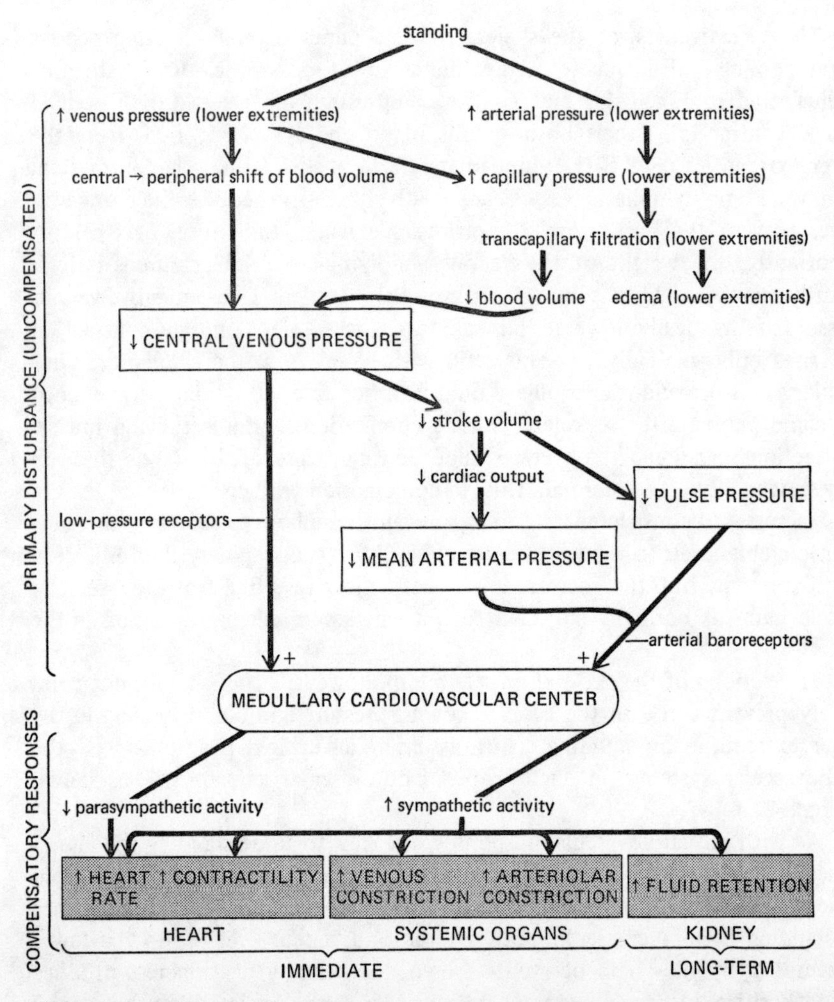

Figure 8-2 Cardiovascular mechanisms involved when changing from a recumbent to a standing position.

vascular variables are not directly influenced by standing but *are* changed by the compensatory responses. Stroke volume and cardiac output, on the other hand, are usually decreased below their recumbent values during quiet standing despite the reflex adjustments that tend to increase them. This is because the positive reflex influences do not quite overcome the negative direct influences on these variables caused by standing. This is in keeping with the general dictum that short-term cardiovascular compensations are never quite complete.

Mean arterial pressure is often found to increase when a person changes from the recumbent to the standing position. At first glance, this is a violation of many rules of cardiovascular system operation. How can a compensation be more than complete? Moreover, how is increased sympathetic activity compatible with higher than normal mean arterial pressure in the first place? In the case of standing, there are many answers to these apparent puzzles. First, average arterial baroreceptor discharge rate can actually decrease in spite of a small increase in mean arterial pressure *if* there is simultaneously a sufficiently large decrease in pulse pressure. Second, mean arterial pressure determined by sphygmomanometry from the arm of a standing individual overestimates the mean arterial pressure actually being sensed by the baroreceptors in the carotid sinus region of the neck. Third, the positive influence on the medullary cardiovascular center from low-pressure receptors may raise the arterial pressure by mechanisms shown in Fig. 7-7B.

The kidney is especially susceptible to changes in sympathetic nerve activity, and consequently, as shown in Fig. 8-2, every reflex alteration in sympathetic activity has influences on fluid balance that become important in the long term. Standing, which is associated with an increase in sympathetic tone, ultimately results in an increase in fluid volume. The ultimate benefit of this is that an increase in blood volume generally reduces the magnitude of the reflex alterations required to tolerate upright posture. (See also study questions 44 to 46.)

EXERCISE

Physical exercise is one of the most ordinary yet taxing situations with which the cardiovascular system must cope. Some of the cardiovascular alterations that occur when a normal person exercises are shown in Fig. 8-3. Note especially that heart rate and cardiac output increase greatly during exercise and that mean arterial pressure and pulse pressure also increase significantly. These alterations ensure that the increased metabolic demands of the exercising skeletal muscle are met by appropriate increases in skeletal muscle blood flow. Many of the adjustments to exercise are due to a large increase in sympathetic activity, which results from the mechanisms outlined in Fig. 8-4.

As indicated among the primary disturbances shown in the top half of Fig. 8-4, the stress (anticipation and/or effort) of exercise exerts a positive influence on the cardiovascular center through corticohypothalamic pathways. This nonbaroreceptor influence in itself will increase sympathetic activity moderately and cause mean arterial pressure to be regulated to a higher than normal level, as discussed in Chap. 7 (see Fig. 7-7A). Also indicated in Fig. 8-4 is the possibility that a second positive nonbaroreceptor influence may reach the cardiovascular center from chemoreceptors in the active skeletal muscles. Such an input would also contribute to the elevations in sympathetic activity and mean arterial pressure that accompany exercise.

The major disturbance on the cardiovascular system during exercise, how-

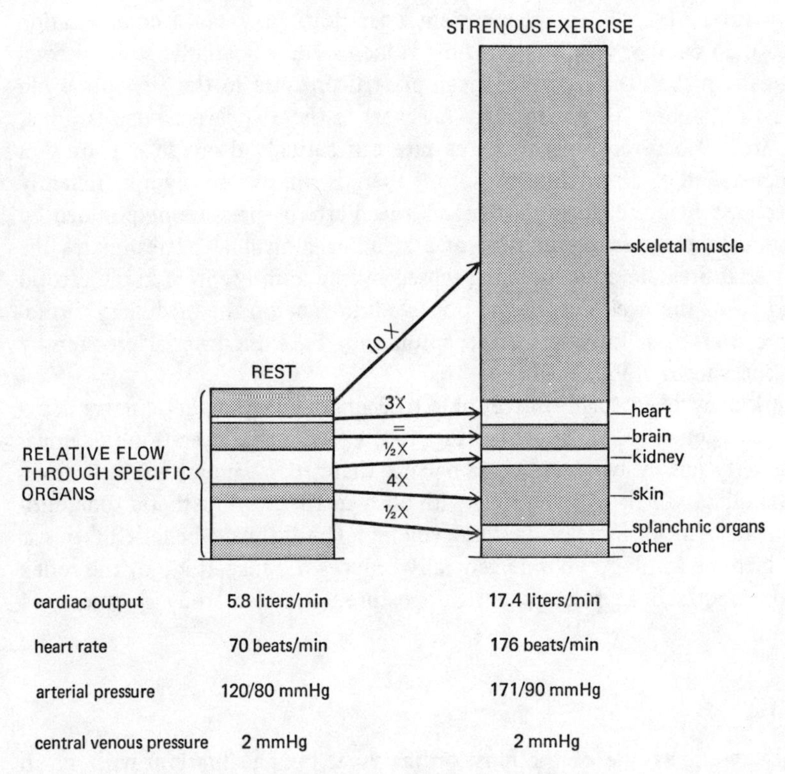

Figure 8-3 Cardiovascular adjustments to strenuous exercise.

ever, is the great decrease in total peripheral resistance caused by metabolic vasodilator accumulation and decreased vascular resistance in active skeletal muscle. By itself, and as indicated in Fig. 8-4, decreased total peripheral resistance is a pressure-lowering disturbance that elicits a strong increase in sympathetic activity through the baroreceptor reflex (see Fig. 7-6B).

Although mean arterial pressure is above normal during exercise the decreased total peripheral resistance causes it to fall below the elevated level to which it would otherwise be regulated by the positive nonbaroreceptor influences on the cardiovascular center. As shown in Fig. 7-7B the baroreceptor reflex pathway responds to this circumstance with a large increase in sympathetic activity. Thus the baroreceptor reflex is responsible for a large portion of the increase in sympathetic activity that accompanies exercise despite the seemingly contradictory fact that arterial pressure is higher than normal. In fact, were it not for the baroreceptor reflex the decrease in total peripheral resistance that occurs during exercise would cause mean arterial pressure to fall well below normal.

As discussed in Chap. 7, and indicated in Figs. 8-3 and 8-4, cutaneous blood flow may increase during exercise despite a generalized increase in sympathetic vasoconstrictor tone because thermal reflexes can override pressure reflexes in the special case of skin blood flow control. Temperature reflexes, of course, are usually activated during strenuous exercise to dissipate the excess heat being produced by the active skeletal muscles. Often cutaneous flow decreases at the onset of exercise (as part of the generalized increase in arteriolar

Figure 8-4 Cardiovascular mechanisms involved during exercise.

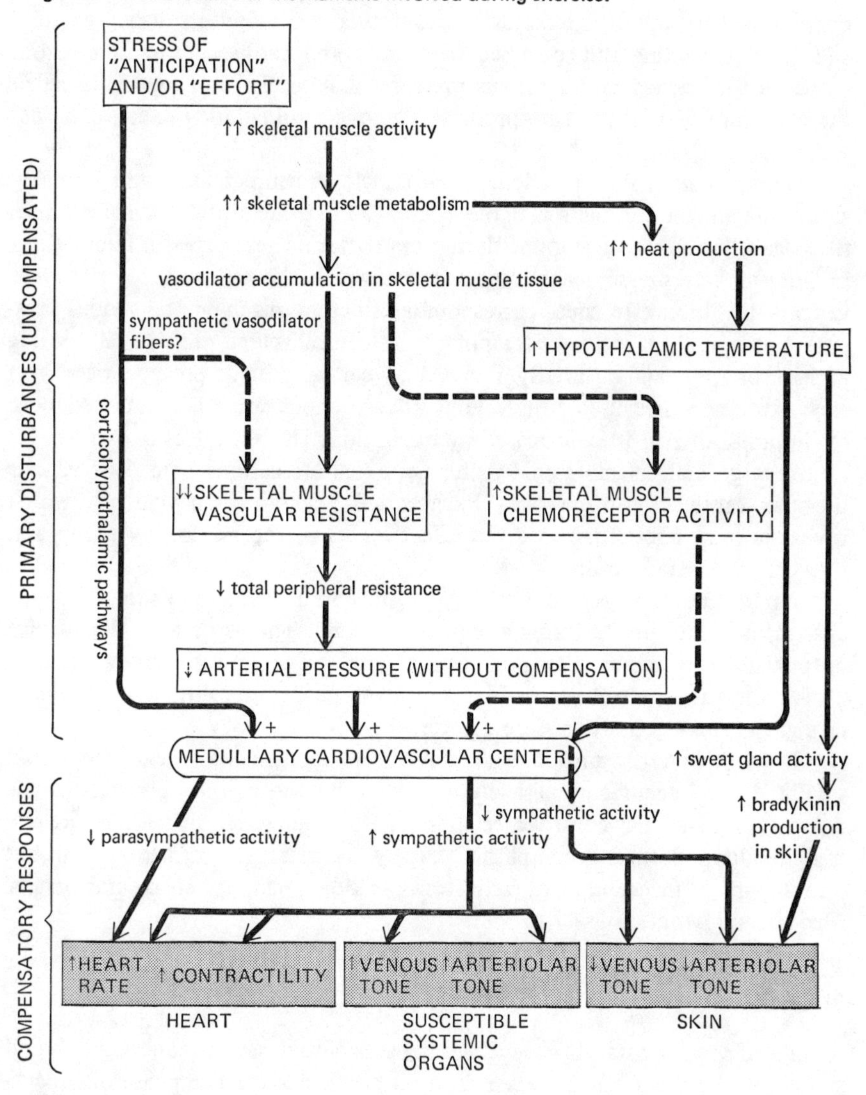

tone from increased sympathetic vasoconstrictor activity) and then increases later during exercise as body heat and temperature build up.

In addition to the increases in skeletal muscle and skin blood flow, coronary blood flow increases substantially during strenuous exercise. This is primarily due to local metabolic vasodilation of coronary arterioles as a result of increased myocardial oxygen consumption.

Two important mechanisms that participate in the cardiovascular response to exercise are not shown in Fig. 8-4. The first is the skeletal muscle pump, which was discussed in connection with upright posture. The skeletal muscle pump is a very important factor in promoting venous return during exercise and thus preventing the increased heart rate and cardiac contractility from drastically lowering central venous pressure. A second factor, which is called the "respiratory pump," also promotes venous return during exercise, as described below.

During a normal inspiration, intrathoracic pressure falls about 7 mmHg as the diaphragm contracts and the chest wall expands. Intrathoracic pressure rises again by an equal amount during expiration. These periodic fluctuations in intrathoracic pressure are transmitted through the thin walls of the great veins in the thorax to cause corresponding fluctuations in central venous pressure. Venous return increases temporarily as central venous pressure falls during each inspiration and is briefly reduced as central venous pressure rises with each expiration. Because of the venous valves, venous return is augmented more by inspiration than it is decreased by expiration. The net effect is that venous return is generally facilitated by the periodic fluctuations in central venous pressure caused by respiration. Exaggerated respiratory movements, which occur during exercise, increase the effectiveness of the respiratory pump and thus enhance venous return.

As indicated in Fig. 8-3, the average central venous pressure does not change much, if at all, during strenuous exercise. This is because the cardiac output and the venous return curves are both shifted upward during exercise. Thus the cardiac output and venous return will be elevated without a significant change in central venous pressure. (Review Fig. 6-5.)

In summary, the profound cardiovascular adjustments to exercise shown in Fig. 8-4 all occur automatically as a consequence of the operation of the normal cardiovascular control mechanisms. The tremendous increase in skeletal muscle blood flow is accomplished largely by increased cardiac output but also in part by diverting flow away from the kidneys and the splanchnic organs. (See also study questions 47 to 52.)

SHOCK

A state of shock exists whenever there is a generalized, severe reduction in blood supply to the body tissues. Even with all cardiovascular compensatory mechanisms activated, arterial pressure is usually low in shock.

In general, shock is precipitated by either severely depressed myocardial functional ability (*cardiogenic shock*) or grossly inadequate cardiac filling. The latter situation can be caused by any number of conditions that decrease central venous volume. *Hemorrhagic shock* accompanies severe blood loss. Severe *burns*, chronic *diarrhea*, or extensive *vomiting* can induce shock by depleting body fluids. Moreover, shock may be associated with widespread loss of arteriolar and venous tone, which can permit peripheral blood pooling. *Anaphylactic shock* occurs as a result of histamine-induced vasodilation in severe allergic reactions. *Septic shock* is caused by vasodilator substances released from infective agents. *Neurogenic shock* is produced by loss of vascular tone due to inhibition of the normal tonic activity of the sympathetic vasoconstrictor nerves and often occurs with deep general anesthesia or in reflex response to the severe deep pain associated with traumatic injuries.

As shown in the top half of Fig. 8-5, the common primary disturbances in all forms of shock are decreased cardiac output and decreased mean arterial pressure. Generally the reduction in arterial pressure is substantial, and so therefore is the positive influence on the cardiovascular center from reduced baroreceptors discharge rate. If arterial pressure falls below about 60 mmHg, brain blood flow begins to fall and this elicits the cerebral ischemic response. As indicated in Chap. 7, the cerebral ischemic response causes the most intense of all activations of sympathetic nerves.

With the exception of neurogenic shock, where reflex responses may be completely absent, shock elicits the compensatory responses we would expect from a fall in blood pressure. These are indicated in the bottom half of Fig. 8-5. The compensatory responses to shock, however, may be much more intense than those that accompany more ordinary cardiovascular disturbances. In addition to increased total peripheral resistance, an important benefit of intense arteriolar constriction during shock is decreased capillary pressure. This causes an increase in blood volume by transcapillary absorption of interstitial fluid from many tissues and also by absorption of fluid from the gastrointestinal tract. Within 30 min after severe hemorrhage, more than 1 liter of fluid may be added to the depleted blood volume by such transcapillary absorption.

Often the strong compensatory responses elicited during shock are capable of preventing drastic reductions in arterial pressure. However, because the compensatory mechanisms involve intense arteriolar vasoconstriction, perfusion of tissues other than the heart and brain may be inadequate despite nearly normal arterial pressure. For example, blood flow through organs such as the liver and kidneys may be reduced nearly to zero by intense sympathetic activation. The possibility of permanent renal or hepatic ischemic damage is a very real concern even in seemingly mild shock situations. Often patients who have apparently recovered from a state of shock die several days later because of renal failure and uremia.

The immediate danger with shock is that it may enter the *progressive stage*, wherein the general cardiovascular situation progressively degenerates, or, worse

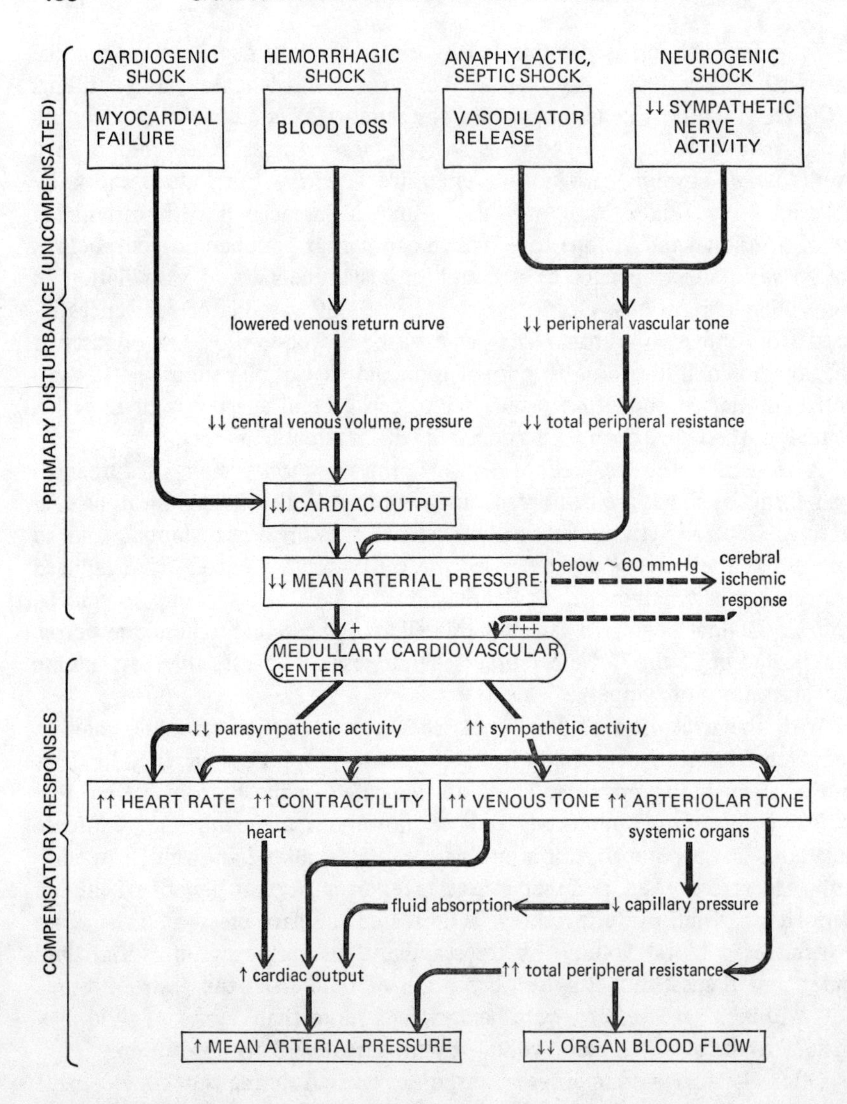

Figure 8-5 Cardiovascular alterations in shock.

yet, enter the *irreversible stage*, where no intervention can halt the ultimate collapse of cardiovascular function that results in death.

The mechanisms behind progressive and irreversible shock are not completely understood. However, it is clear, as shown in Fig. 8-6, that bodily homeostasis progressively deteriorates with prolonged reductions in organ blood flow. These homeostatic disturbances in turn adversely affect various components of the cardiovascular system so that arterial pressure and thus organ blood flow is further reduced. Note that the events shown in Fig. 8-6 are *decompensatory mechanisms.* Reduced arterial pressure leads to alterations that further reduce arterial pressure rather than tend to correct it. If the shock state is severe enough and/or has persisted long enough to enter the progressive stage, the self-reinforcing decompensatory mechanisms progressively drive arterial pressure down. Unless corrective measures are taken quickly, death will ultimately result. (See also study questions 53 and 54.)

CHRONIC HEART FAILURE

Heart (or cardiac, or myocardial) *failure* is said to exist whenever ventricular function is depressed through myocardial damage, insufficient coronary flow,

Figure 8-6 Decompensatory mechanisms in shock.

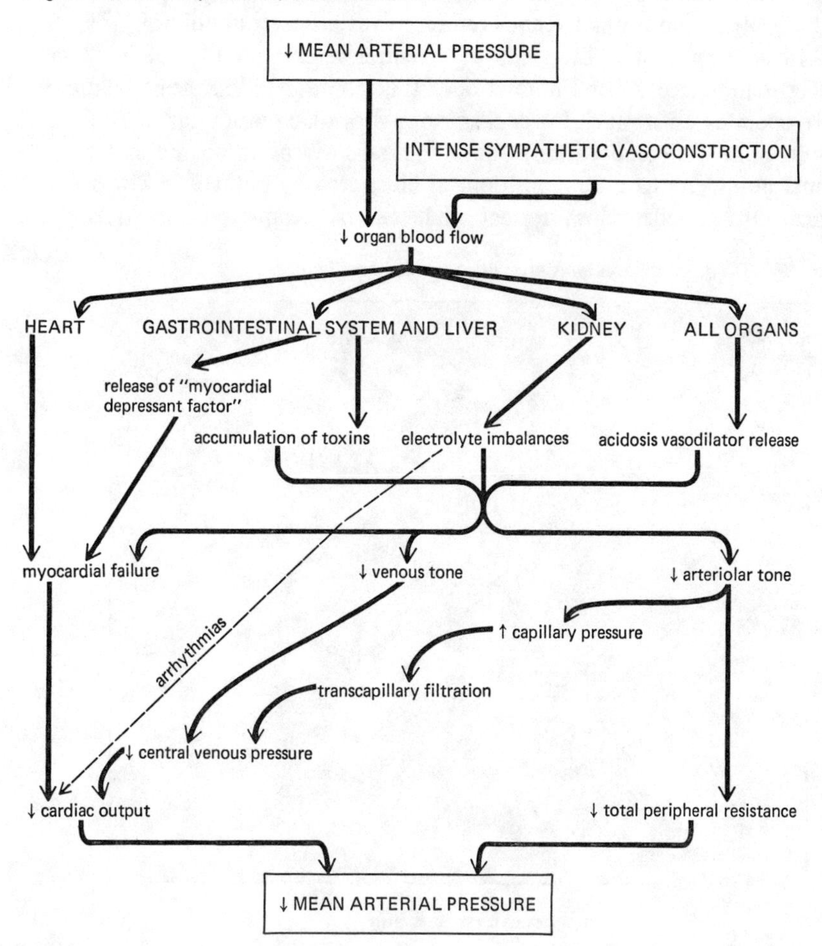

or any other condition that directly impairs the mechanical performance of heart muscle. By definition, heart failure implies a *lower than normal cardiac function curve.* We have already discussed acute heart failure in the context of cardiogenic shock and as part of the decompensatory mechanisms operating in progressive and irreversible shock. Often, however, conditions such as progressive coronary artery disease may induce a chronic state of heart failure.

The primary disturbance in heart failure (acute or chronic) is depressed cardiac output and thus lowered arterial pressure. Consequently, all the compensatory responses important in shock (Fig. 8-5) are also important in heart failure. In chronic heart failure, however, the cardiovascular disturbances may not be sufficient to produce a state of shock. Moreover, long-term compensatory mechanisms are especially important in chronic heart failure.

The circumstances of chronic heart failure are well illustrated by cardiac output and venous return curves such as those shown in Fig. 8-7. The normal cardiac output and normal venous return curves intersect at point A in Fig. 8-7. A cardiac output of 5 liters/min at a central venous pressure of less than 2 mmHg is indicated by the normal operating point (A). With heart failure, the heart operates on a much lower than normal cardiac output curve. Thus heart failure alone (uncompensated) shifts the cardiovascular operation from the normal point (A) to a new position, as illustrated by point B in Fig. 8-7; i.e., cardiac output falls below normal while central venous pressure rises above

Figure 8-7 Cardiovascular alterations with chronic heart failure.

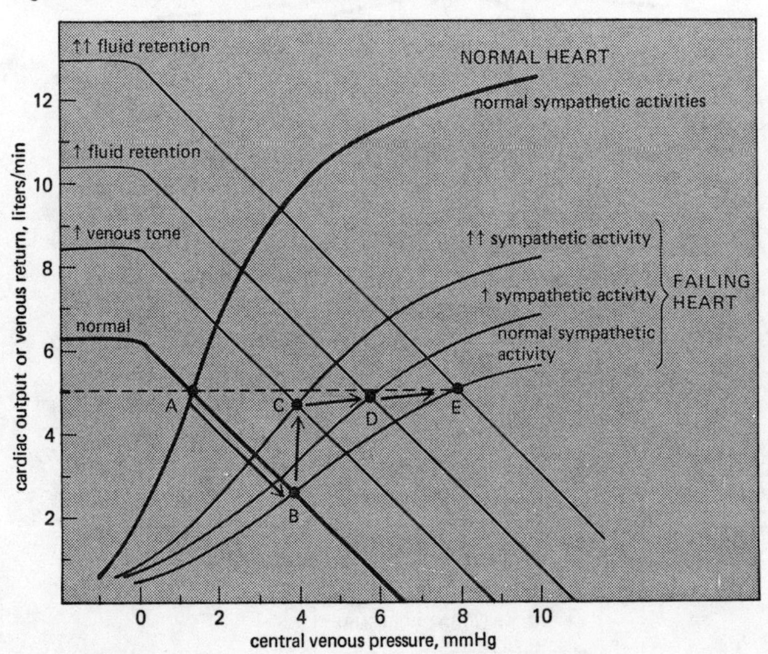

normal. The decreased cardiac output leads to decreased arterial pressure and reflex activation of the cardiovascular sympathetic nerves. Increased sympathetic nerve activity tends to (1) raise the cardiac output curve toward normal and (2) increase peripheral venous pressure through venous constriction and thus raise the venous return curve above normal. Cardiovascular operation will shift from point B to point C in Fig. 8-7. Thus the depressed cardiac output is substantially improved by the immediate consequences of increased sympathetic nerve activity. Note, however, that the cardiac output at point C is still below normal. The arterial pressure associated with cardiovascular operation at point C is likely to be near normal, however, since higher than normal total peripheral resistance will accompany higher than normal sympathetic nerve activity.

In the long term, cardiovascular operation cannot remain at point C in Fig. 8-7. Operation at point C involves higher than normal sympathetic activity, and this will inevitably cause a gradual increase in blood volume by the mechanisms that were described in Chap. 7 and were indicated in Fig. 8-2 (lower right). Over several days, there is a progressive rise in the venous return curve as a result of increased blood volume. This will shift the cardiovascular operating point from C to D to E as shown in Fig. 8-7.

Note that increased fluid retention (C → D → E in Fig. 8-7) causes a progressive increase in cardiac output toward normal and simultaneously allows a reduction in sympathetic nerve activity toward the normal value. Reduced sympathetic activity is beneficial for several reasons. First, decreased arteriolar constriction permits renal and splanchnic blood flow to increase toward more normal values. Second, myocardial oxygen consumption may fall as sympathetic nerve activity falls, even though cardiac output tends to increase. Recall that increased heart rate and increased cardiac contractility greatly increase myocardial oxygen consumption. Reduced myocardial oxygen consumption is especially beneficial in situations where inadequate coronary blood flow is the cause of the heart failure.

Unfortunately, the consequences of fluid retention in cardiac failures are not all beneficial. Note in Fig. 8-7 that fluid retention (C → D → E) will cause both peripheral and central venous pressures to be much higher than their normal values. Chronically high central venous pressure causes chronically increased end-diastolic volume (cardiac dilation). Up to a point, cardiac performance is improved by increased cardiac filling volume through Starling's law. Excessive cardiac dilation, however, can impair cardiac function because increased wall tension is required to generate pressure within an enlarged ventricular chamber ($T = Pr$, Chap. 2).

The high venous pressure associated with fluid retention also adversely affects organ function because transcapillary fluid filtration, edema formation, and congestion are produced by a high venous pressure. Pulmonary edema and respiratory crisis often accompany left heart failure. Common signs of right heart failure include distended neck veins, peripheral edema, and fluid accumulation in the abdomen (ascites) with liver congestion and dysfunction.

In the example shown in Fig. 8-7, the depresssion in the cardiac output curve due to heart failure is only moderately severe. Thus it is possible, through moderate fluid retention, to achieve a normal cardiac output with essentially normal sympathetic activity (point E). The situation at point E is relatively stable because the stimuli for further fluid retention have been removed. If, however, the heart failure is more severe, the cardiac output curve may be so depressed that normal cardiac output cannot be achieved by any amount of fluid retention. In these cases fluid retention is extremely marked, as is the elevation in venous pressure, and the adverse complications of congestion are very serious problems. Usually, cardiac glycosides such as digitalis are used in the treatment of severe congestive heart failure in an attempt to raise the cardiac output curve by pharmacological means. (See also study questions 54 to 57.)

HYPERTENSION

Hypertension (chronic high arterial pressure) is an extremely common cardiovascular problem. It has been established beyond doubt that hypertension increases the risk of coronary artery disease, myocardial infarction, stroke, and many other serious cardiovascular problems. Moreover, it has been clearly demonstrated that the risk of serious cardiovascular incidents is reduced by proper treatment of hypertension.

In approximately 90 percent of cases the primary abnormality that produces high blood pressure is unknown. In the remaining 10 percent of hypertensive patients, the cause can be traced to epinephrine-producing tumors (pheochromocytomas), aldosterone-producing tumors (in primary hyperaldosteronism), or certain forms of renal disease (renal artery stenosis, glomerular nephritis, toxemia of pregnancy). Most often, however, the true cause of the hypertension remains a mystery and it is only the symptom of high blood pressure that is treated. The term *essential hypertension* is applied to this situation.

As discussed in Chap. 7, the urine output rate is influenced by arterial pressure, and in the long term arterial pressure can stabilize only at the level that makes urine output rate equal to fluid intake rate. As shown by point N in Fig. 8-8, this pressure is approximately 100 mmHg in a normal individual.

All forms of hypertension involve an alteration somewhere in the chain of events through which changes in arterial pressure produce changes in urine output rate (see Fig. 7-9) such that the renal function curve is shifted rightward as indicated in Fig. 8-8. The important feature to note is that *higher than normal arterial pressure is required to produce a normal urine output rate in hypertension.*

Consider that the untreated hypertensive individual in Fig. 8-8 would have

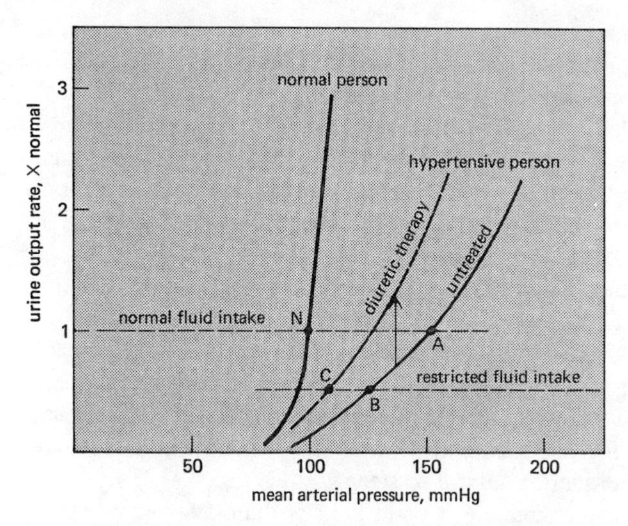

Figure 8-8 Renal function curves in hypertension and hypertension therapy.

a very low urine output rate at the normal mean arterial pressure of 100 mmHg. Recall from Fig. 7-8 that whenever the fluid intake rate exceeds the urine output rate, fluid volume must rise and consequently so will cardiac output and mean arterial pressure. With a normal fluid intake rate, this untreated hypertensive patient will ultimately stabilize at point A (mean arterial pressure = 150 mmHg). Recall from Chap. 7 that the baroreceptors adapt within days so that they have a normal discharge rate at the *prevailing* average arterial pressure. Thus, once the hypertensive individual has been at point A for a week or more, even the baroreceptor mechanism will begin resisting acute changes from the 150-mmHg pressure level.

A mean arterial pressure of 150 mmHg obviously must be ultimately produced by either higher than normal total peripheral resistance or higher than normal cardiac output ($\bar{P}_A = \text{CO} \times \text{TPR}$). From the reasoning in the preceding paragraph, we might well expect hypertension always to be accompanied by higher than normal blood volume and higher than normal cardiac output. Most often, however, neither blood volume nor cardiac output is found to be higher than normal in patients with well-established hypertension; instead, it is the total peripheral resistance that is substantially increased. The mechanisms that cause the increase in total peripheral resistance that frequently accompanies hypertension are not completely understood. One possibility is that the simultaneous autoregulation of blood flow in many organs (Chap. 5) results in "total body autoregulation." In this case, total peripheral resistance increases (by mechanisms discussed in Chap. 5) so that total systemic flow (i.e., cardiac output) will remain constant in the presence of increased mean arterial pressure.

Also, in many cases of hypertension, blood levels of renin and angiotensin II are high. In these cases, increased total peripheral resistance may well be caused by the powerful vasoconstrictor action of angiotensin II.

A most important fact to realize is that, although high blood pressure must always ultimately be sustained by either high cardiac output or high total peripheral resistance, neither need be the primary cause. A shift in the relationship between arterial pressure and urine output rate, as illustrated in Fig. 8-8, however, will always produce hypertension.

In certain hypertensive individuals, restricting salt (and therefore fluid) intake produces a substantial reduction in blood pressure. In the example of Fig. 8-8, this effect is illustrated by a shift from point A to point B. The efficacy of lowering fluid intake to lower arterial pressure depends heavily on the slope of the renal function curve in the hypertensive individual. The arterial pressure of a normal individual, for example, is little affected by changes in fluid intake because the normal renal function curve is so steep.

A second common treatment of hypertension is diuretic therapy. Many diuretic drugs are available, but most have the effect of inhibiting renal tubular salt (and therefore fluid) reabsorption. The net effect of diuretic therapy, as shown in Fig. 8-8, is that the urine output rate for a given arterial pressure is increased; i.e., diuretic therapy raises the renal function curve. The combined result of restricted fluid intake and diuretic therapy for the hypertensive individual of Fig. 8-8 is illustrated by point C. (See also study questions 58 and 59.)

Study questions: **44** to **59**

Study Questions

QUESTIONS

ANSWERS

1 Whenever skeletal muscle blood flow increases, blood flow to other organs must decrease. True or false?

False. Flow through any vascular bed depends on its resistance to flow and the arterial pressure. As long as this pressure is maintained constant (a critical point), alterations in flow through any individual bed will have no influence on flow through other beds in parallel with it.

2 *a* Determine the vascular resistance of a resting skeletal muscle from the following data:

Mean arterial pressure = 100 mmHg

a. Since

$$\dot{Q} = \frac{\Delta P}{R}$$

Mean venous pressure = 0 mmHg
Blood flow to the muscle
= 5 ml/min

then

$$R = \frac{\Delta P}{\dot{Q}}$$

Therefore

$$R = \frac{100 \text{ mmHg}}{5 \text{ ml/min}}$$

$$= 20 \text{ mmHg·min/ml}$$

b Assume that when the muscle is exercising, the resistance vessels dilate so that the internal radius doubles. If blood pressure does not change, what is the blood flow to the exercising muscle?

b. According to Poiseuille's equation,

$$\dot{Q} = \Delta P \frac{\pi r^4}{8L} \frac{1}{\eta}$$

With other factors constant,

$$\dot{Q} \propto r^4$$

Thus doubling the radius with exercise increases flow 16-fold over that at rest. Therefore

$$\dot{Q} = 16 \times 5 \text{ ml/min}$$

$$= 80 \text{ ml/min during exercise}$$

c What is the vascular resistance of this exercising skeletal muscle?

c. Since

$$R = \frac{\Delta P}{\dot{Q}}$$

we have

$$R = \frac{100 \text{ mmHg}}{80 \text{ ml/min}}$$

$$= 1.25 \text{ mmHg·min/ml}$$

3 Calculate the cardiac output from the following data:

Pulmonary arterial pressure
= 20 mmHg

The blood flow rate through the lungs ($\dot{Q}_L$) must equal the cardiac output (CO) because of the way the cardiovascular system is arranged. Thus

Pulmonary venous pressure
= 0 mmHg
Pulmonary vascular resistance
= 4 mmHg·min/liter

$$CO = \dot{Q}_L = \frac{\Delta P_L}{R_L}$$

$$= \frac{20 \text{ mmHg}}{4 \text{ mmHg·min/liter}}$$

$$= 5 \text{ liter/min}$$

4 Determine the rate of glucose uptake by an exercising skeletal muscle ($\dot{G}_m$) from the following data:

Arterial blood glucose concentration,

$[G]_a = 50$ mg per 100 ml

Muscle venous blood glucose concentration,

$[G]_v = 30$ mg per 100 ml

Blood flow,

$\dot{Q} = 60$ ml/min

The Fick principle states that

$$\dot{G}_m = \dot{Q}([G]_a - [G]_v)$$

Thus

$$\dot{G}_m = 60 \text{ ml/min} \times \frac{(50-30) \text{ mg}}{100 \text{ ml}}$$

$$= 12 \text{ mg/min}$$

5 Determine the direction of transcapillary fluid movement ($\dot{F}$) within a tissue, given the following data:

Capillary hydrostatic pressure,

$P_c = 28$ mmHg

Plasma oncotic pressure,

$\pi_c = 24$ mmHg

Tissue hydrostatic pressure,

$P_i = -4$ mmHg

Tissue oncotic pressure,

$\pi_i = 0$ mmHg

Since

$$\dot{F} = K[(P_c - P_i) - (\pi_c - \pi_i)]$$

then

$$\dot{F} = K[28 - (-4) - 24 + 0] \text{ mmHg}$$

$$= K \times 8 \text{ mmHg}$$

Since this is a positive term, the net rate of filtration ($\dot{F}$) is positive, indicating net movement of fluid out of the capillaries.

6 Which of the following conditions favor edema formation?

a Lymphatic blockage
b Thrombophlebitis (venous clot)

All do: *a* and *d* by allowing interstitial protein buildup, *b* by raising P_c, and *c* for obvious reasons.

c Decreased plasma protein concentration

d Greatly increased capillary pore size

7 *a* What will happen to the potassium equilibrium potential of cardiac muscle cells when interstitial $[K^+]$ is elevated?

a. The potassium equilibrium potential will decrease because less potential difference is required to balance the decreased tendency for net K^+ diffusion out of the cell. [$E_{eq\,K^+} = (-61.5\ \text{mV}) \log([K^+]_i/[K^+]_o)$.]

b What effect will this have on the cells' resting membrane potential?

b. Since the resting membrane is most permeable to K^+, the resting membrane potential is always close to the K^+ equilibrium potential. Lowering the K^+ equilibrium potential will undoubtedly also lower the resting membrane potential.

8 A decrease in AV nodal conduction velocity will

- *a* Decrease heart rate
- *b* Increase P wave amplitude
- *c* Increase the PR interval
- *d* Widen the QRS complex
- *e* Increase ST segment duration

Only *c*.

9 If the R wave is upright and equally large on leads II and III, what is the mean electrical axis of the heart?

According to the electrocardiographic conventions, the electrical axis is 90°. Furthermore, the R wave will not appear in lead I recordings since the electrical dipole at this instant is perpendicular to the lead I axis.

10 List some of the events of the cardiac cycle associated with the phase of isovolumic contraction.

(1) QRS complex of the electrocardiogram immediately precedes ventricular contraction.

(2) AV values close at the beginning of this phase.

(3) Intraventricular pressure rises rapidly.

(4) Aortic pressure reaches its lowest value.

(5) First heart sound occurs.

(6) Aortic and pulmonic valves open at the end of the interval.

11 If pulmonary artery pressure is 24/8 (systolic/diastolic), what are the respective systolic and diastolic pressures of the right ventricle?

The ventricular systolic pressure is also 24 mmHg since the normal pulmonic valve provides negligible resistance to flow during ejection. The right ventricular diastolic pressure, however, is determined by systemic venous pressure and will be close to 0 mmHg.

12 Because pulmonary artery pressure is so much lower than aortic pressure, the right ventricle has a larger stroke volume than the left ventricle. True or false?

False. Although there may be minor beat-to-beat inequalities, the average stroke volumes of the right and left ventricles must be equal or blood would accumulate in the pulmonic or systemic circulation.

13 Which of the following arrhythmias might result in a reduced stroke volume?

a Paroxysmal atrial tachycardia
b Ventricular tachycardia
c Atrial fibrillation
d Ventricular fibrillation
e Third-degree heart block

a and *b*, because filling time is reduced; *c*, if ventricular rate is rapid; *d*, for obvious reasons; but not *e*, because ventricular pacemakers produce a lower heart rate, which is usually associated with a larger stroke volume.

14 Describe the primary pressure abnormalities associated with

a Aortic stenosis
b Mitral stenosis

a. Aortic stenosis produces a significant pressure difference between the left ventricle and the aorta during systolic ejection.
b. Mitral stenosis produces a significant pressure difference between the left atrium and the left ventricle during diastole.

15 If the left ventricular chamber is enlarged, the wall tension required to generate a given systolic pressure is increased. True or false?

True. The law of Laplace states that when the radius (r) of a sphere increases, the wall tension (T) for a given internal pressure (P) must also increase:

$$T = Pr$$

16 Which of the following interventions will increase cardiac stroke volume?

a Increased ventricular filling pressure

All of them: *a* by increasing preload, *b* by decreasing afterload, and *c* and *d* by augmenting contractility.

b Decreased arterial pressure
c Increased activity of cardiac sympathetic nerves
d Increased circulating catecholamine levels

17 Given the following information, calculate cardiac output:

Systemic arterial blood O_2 concentration,

$$[O_2]_{SA} = 200 \text{ ml/liter}$$

Pulmonary arterial blood O_2 concentration,

$$[O_2]_{PA} = 140 \text{ ml/liter}$$

$$O_2 \text{ consumption} = 600 \text{ ml/min}$$

$$\dot{Q} = \frac{O_2 \text{ consumption}}{[O_2]_{SA} - [O_2]_{PA}}$$

$$= \frac{600 \text{ ml/min}}{(200 - 140) \text{ ml/liter}}$$

$$= 10 \text{ liters/min}$$

18 In which direction will cardiac output change if central venous pressure is lowered while cardiac sympathetic tone is increased?

One cannot tell from the information given because the two alterations would have opposite effects on cardiac output. A complete set of ventricular function curves, as well as quantitative information about the changes in filling pressure and sympathetic tone, would be necessary to answer the question.

19 Given the following data, calculate an individual's total peripheral resistance:

Mean arterial pressure,

$$\bar{P}_A = 100 \text{ mmHg}$$

Central venous pressure,

$$P_{CV} = 0 \text{ mmHg}$$

Cardiac output,

$$CO = 6 \text{ liters/min}$$

Since

$$\dot{Q} = \frac{\Delta P}{R}$$

then

$$R = \frac{\Delta P}{\dot{Q}}$$

and

$$TPR = \frac{\bar{P}_A - P_{CV}}{CO}$$

Therefore

$$TPR = \frac{(100-0)\ \text{mmHg}}{6\ \text{liters/min}}$$

$$= 16.7\ \text{mmHg} \cdot \text{min/liter}$$

20 The total peripheral resistance to blood flow is greater than the resistance to flow through any of the systemic organs. True or false?

False. It is less than the resistance to flow through any of the organs. Each organ, in effect, provides an additional pathway through which blood may flow; thus the individual organ resistances must be greater than the total resistance and

$$\frac{1}{TPR} = \frac{1}{R_1} + \frac{1}{R_2} + \cdots + \frac{1}{R_n}$$

21 Decreasing the renal vascular resistance will increase TPR. True or false?

False. Since

$$\frac{1}{TPR} = \frac{1}{R_{\text{kidneys}}} + \cdots$$

a decrease in renal resistance must increase 1/TPR and therefore decrease TPR. When the resistance of any single peripheral organ changes, TPR changes in the same direction.

22 Is it possible for total arterial blood flow to be greater than total capillary blood flow?

No. If flow through arteries exceeded flow through capillaries for any significant period of time, arterioles would explode.

23 Constriction of arterioles in an organ promotes reabsorption of interstitial fluid from that organ. True or false?

True. Since arteriolar constriction tends to reduce the hydrostatic pressure in the capillaries, reabsorptive forces will exceed filtration forces and net reabsorption of interstitial fluid into the vascular bed will occur.

24 Which of the following increase blood flow through a skeletal muscle?

- *a* Increase in tissue P_{CO_2}
- *b* Increase in tissue adenosine
- *c* Alpha-receptor blocking drugs
- *d* Sympathetic stimulation

a, b, and *c.*

25 Autoregulation of blood flow implies that arterial pressure is adjusted by local mechanisms to ensure constant flow through an organ. True or false?

False. Autoregulation of blood flow implies that vascular resistance is adjusted to maintain constant flow in spite of changes in arterial pressure.

26 Coronary blood flow will normally increase when

a Arterial pressure increases
b Heart rate increases
c Sympathetic activity increases
d The heart is dilated

All, primarily because all increase myocardial oxygen consumption.

27 Blood flow through organs containing arterioles with high intrinsic tone is controlled primarily by neurogenic mechanisms. True or false?

False. Flow through these organs is primarily determined by local metabolic vasodilator mechanisms. Neurogenic control is dominant in organs that have blood flows far in excess of that required to meet metabolic needs–that is, in organs with low intrinsic vascular tone.

28 A person who hyperventilates (breathes rapidly and deeply) gets dizzy. Why?

Hyperventilation decreases the blood P_{CO_2} level. This, in turn, causes cerebral arterioles to constrict (recall that cerebral vascular tone is highly sensitive to changes in P_{CO_2}). The increased cerebral vascular resistance causes a decrease in cerebral blood flow, which produces dizziness and disorientation.

29 A patient complains of severe leg pains after walking a short distance. The pains disappear after the patient rests (this symptom is called *intermittent claudication*). What might be the problem?

It is likely that the increased metabolic demands evoked by the exercising skeletal muscle cannot be met by an appropriate increase in blood flow to the muscle. This patient may have some sort of arterial disease (atherosclerosis) that provides a high resistance to flow that cannot be overcome by local metabolic vasodilator mechanisms.

30 How would a stenotic aortic valve influence coronary blood flow?

High left ventricular pressures must be developed to eject blood through the stenotic valve (Fig. 2-12). This increases myocardial oxygen consumption, which tends to increase coronary flow. At the same time, however, high intraventricular pressure development enhances the systolic compression of coronary vessels and tends to decrease flow. Coronary perfusion pressure will also be decreased if the systemic arterial pressure is lower than normal.

31 What determines central venous pressure?

Central venous pressure always settles at the value that makes cardiac output and venous return equal. Therefore anything that shifts the cardiac function curve or the venous return curve affects venous pressure.

32 According to Starling's law, cardiac output always decreases whenever central venous pressure decreases. True or false?

False. Starling's law says that, *if other influences on the heart are constant*, cardiac output decreases when central venous pressure decreases (e.g., A → B in Fig. 6-6). In the intact cardiovascular system, where many things may happen simultaneously, cardiac output and central venous pressure may change in opposite directions (e.g., B → C in Fig. 6-6).

33 Venous return will be greater than cardiac output when

a Peripheral venous pressure is higher than normal

b Blood volume is higher than normal

c Cardiac sympathatic nerve activity is lower than normal

None. Venous return must always equal cardiac output in an equilibrium situation.

34 Chronic elevation of arterial pressure requires that either cardiac output or total peripheral resistance (or both) be chronically elevated. True or false?

True. $\bar{P}_A = CO \times TPR$.

35 Whenever cardiac output is increased, mean arterial pressure *must* also be increased. True or false?

False. Increases in cardiac output are often accompanied by decreases in total peripheral resistance so that mean arterial pressure is maintained constant.

36 Acute increases in arterial pulse pressure usually result from increases in stroke volume. True or false?

True. $P_p \simeq SV/C_A$. Acute changes in arterial compliance usually do not occur.

37 An increase in total peripheral resistance increases diastolic pressure (P_D) more than systolic pressure (P_S). True or false?

False. Changes in TPR (with CO constant) produce approximately equal increases in P_S and P_D and increase $\bar{P}_A$ with little influence on pulse pressure.

38 Estimate the mean arterial pressure when the measured arterial pressure is 110/70 mmHg.

$$\begin{aligned}\bar{P}_A &= P_D + \tfrac{1}{3}(P_S - P_D)\\ &= 70 + \tfrac{1}{3}(110 - 70)\text{ mmHg}\\ &= 83\text{ mmHg}\end{aligned}$$

39 Consider the various components of the baroreceptor reflex and predict whether the following variables will increase or decrease in response to a *rise* in arterial pressure.

a Baroreceptor firing rate
b Tonic activity of the vasomotor center
c Tonic activity of the cardioinhibitory center
d Parasympathetic activity to the heart
e Sympathetic activity to the heart
f Arteriolar tone
g Venous tone
h Peripheral venous pressure
i Total peripheral resistance
j Cardiac output

a, c, and *d* will increase; the rest will decrease.

40 Massage of the neck over the carotid sinus area in a person experiencing a bout of paroxysmal atrial tachycardia is often effective in terminating the episode. Why?

Carotid sinus massage causes baroreceptors to fire, which in turn increases the activity of the cardioinhibitory center. The increased parasympathetic activity acts to slow the pacemaker activity and allows a more normal rhythm to be established.

41 Indicate whether mean arterial pressure is *increased* or *decreased* when the following receptors are activated:

a Central chemoreceptors
b Peripheral chemoreceptors
c Atrial and pulmonary low-pressure receptors
d Cutaneous pain receptors
e Deep pain receptors

a, b, and *d* increase mean arterial pressure; *c* and *e* decrease mean arterial pressure.

42 Describe the immediate cardiovascular consequences of giving a normal person a drug that blocks alpha-adrenergic receptors.

(1) The influence of sympathetic nerve activity on arteriolar tone will be blocked. Arteriolar tone will fall and thus so will TPR. Alpha blockade represents a pressure-lowering disturbance on the effector portion of the cardiovascular system.
(2) The effector portion function curve will shift downward as shown in Fig. 7-6B. (In this instance the effector function curve may also become less steep because increases in TPR no longer aid in the production of increased $\overline{P}_A$ when sympathetic activity increases.)
(3) A new equilibrium will be established within the baroreceptor reflex pathway at lower than normal arterial pressure and higher than normal sympathetic nerve activity, as shown in Fig. 7-6B.
(4) Heart rate and cardiac output will increase because of the increased sympathetic activity. The cardiac function curve will shift upward, but the venous return curve will not because alpha-receptor blockade blocks the effect of increased sympathetic activity on the veins. Consequently, central venous pressure will be lower than normal (see Fig. 6-5).

43 What long-term adjustments will alpha-receptor blockade cause?

(1) Alpha-receptor blockade will lower the normally high tone of renal arterioles. This will cause higher than normal glomerular capillary pressure, glomerular filtration rate, and thus urine output rate at normal arterial pressure; the renal function curve (Fig. 7-10) will be shifted to the left.
(2) The altered renal function will promote fluid loss until a new equilibrium is reached between fluid intake rate and urine output rate at lower than normal arterial pressure. Thus mean arterial pressure will be chronically lowered by chronic alpha-receptor blockade.

44 How are the thin-walled capillaries able to withstand pressures greater than 100 mmHg without rupturing?

Because capillaries have such a small radius, the tension in the capillary wall is rather modest despite very high internal pressures ($T = Pr$).

45 Soldiers faint when standing at attention on a very hot day more often than on a cooler day. Why?

Fainting occurs because of decreased cerebral blood flow when mean arterial pressure falls below about 60 mmHg. On a hot day, temperature reflexes override pressure reflexes to produce the increased skin blood flow required for thermal regulation. Thus TPR is lower when standing on a hot day than on a cool one. Consequently, mean arterial pressure falls below 60 mmHg with less lowering of cardiac output on a warm day than on a cool one.

46 For several days after an extended period of bed rest, patients often become dizzy when they stand upright quickly because of an exaggerated transient fall in arterial pressure (*orthostatic hypotension*). Why might this be so?

The cardiovascular response to lying down is just the opposite of that shown in Fig. 8-2. Patients tend to lose rather than retain fluid during extended bed rest and end up with lower than normal blood volumes. Thus they are less able to cope with an upright posture during the period required for blood volume to reachieve the value it has when periods of standing are part of the patient's normal routine.

47 How is the decrease in skeletal muscle vascular resistance evident from Fig. 8-3?

$R = \bar{P}_A/\dot{Q}$. Skeletal muscle resistance must have decreased considerably during exercise because skeletal muscle flow increased 10-fold (1000 percent) whereas arterial pressure increased much less ($\simeq$ 25 percent).

48 Is a decrease in total peripheral resistance implied in Fig. 8-3?

$TPR = \bar{P}_A/CO$. Total peripheral resistance must have decreased during exercise because cardiac output increased threefold, which is relatively much larger than the increase in arterial pressure.

49 What in Fig. 8-3 implies increased sympathetic activity?

(1) Decreased renal and splanchnic blood flows in spite of increased mean arterial pressure indicate sympathetic vasoconstriction (Chap. 5).
(2) Increased cardiac output at constant central venous pressure indicates increased cardiac contractility and thus increased activity of cardiac sympathetic nerves (Chap. 3).
(3) The heart rate during exercise is well above the intrinsic rate ($\simeq$ 100 beats per minute). This indicates activation of the cardiac sympathetic nerves because withdrawal of cardiac parasympathetic activity cannot increase heart rate above the intrinsic rate (Chap. 3).

50 What factors cause an increase in myocardial metabolism during exercise?

Myocardial oxygen consumption is increased by increased cardiac external work per minute ($W = \bar{P}_A \times CO$), increased heart rate, and increased contractility (Chap. 7).

51 Most artificial respirators force air into the lungs with positive pressure. Respiration with such a device

When the lungs are inflated artificially, intrathoracic pressure goes up (rather than down, as occurs during normal often produces cardiovascular distress. Why?

inspiration). On the average, intrathoracic pressure and thus central venous pressure are higher than normal with artificial respiration. In this situation, however, higher than normal central venous pressure does not increase cardiac filling significantly because a parallel increase in pressure occurs on the outside of the heart. The increased central venous pressure does inhibit venous return, and this is what causes the adverse cardiovascular effects of positive pressure respiration.

52 Blood pressure can rise to extremely high levels during strenuous isometric exercise maneuvers like weight lifting. Why?

Blood flow through muscle is reduced or stopped by compressive forces on vessels during an isometric muscle contraction. Thus, during an isometric maneuver, TPR may be higher than normal rather than much lower than normal as it is during phasic exercises like running. In the absence of decreased TPR but the presence of strong nonbaroreceptor positive influences on the cardiovascular center, mean arterial pressure may be regulated to very high values (see point 2 in Fig. 7-7B).

53 Clinical signs of shock often include pale and cold skin, dry mucous membranes, weak but rapid pulse, and muscle weakness and mental disorientation or unconsciousness. What are the physiological conditions that account for these signs?

Intense sympathetic activation drastically reduces skin blood flow, promotes transcapillary reabsorption of fluids, stimulates the heart (which still will have a low stroke volume because of low central venous pressure), and reduces skeletal muscle blood flow. Cerebral blood flow falls if the compensatory mechanisms do not prevent mean arterial pressure from falling below 60 mmHg.

54 Which of the following would be helpful to hemorrhagic shock victims?

- ***a*** Keep them on their feet
- ***b*** Warm them up

a. Not helpful since gravity tends to promote peripheral venous blood pooling and cause a further fall in arterial pressure.

c Give them fluids to drink
d Maintain their blood pressure with catecholamine-type drugs

b. Not helpful if carried to an extreme. Cutaneous vasodilation produced by warming adds to the cardiovascular stresses.
c. Helpful if the victim is conscious and can drink since fluid will be rapidly absorbed from the gut to increase circulating blood volume.
d. Might be helpful as an initial emergency measure to prevent brain damage due to severely reduced blood pressure, but prolonged treatment will promote the decompensatory mechanisms associated with decreased organ blood flow.

55 Why are diuretic drugs (see hypertension section) often helpful in treating patients in congestive heart failure?

Excessive fluid retention can induce decompensatory mechanisms that further compromise an already weakened heart (e.g., inadequate oxygenation of the blood as it passes through edematous lungs, marked cardiac dilation and increased myocardial metabolic needs, liver dysfunction due to congestion). Diuretic therapy reduces fluid volume and the high venous pressures that are the cause of these problems.

56 What is the potential danger of vigorous diuretic therapy for the patient in heart failure?

If blood volume and central venous pressure are reduced too far with diuretic therapy, cardiac output may fall to unacceptably low levels through Starling's law.

57 Will an increase in left atrial pressure to 10 mmHg cause pulmonary edema?

Probably not. Recall that there is normally a reabsorptive force of about 17 mmHg across the lung capillaries that keep the lung tissues dry. Thus, even with a 10-mmHg rise in pulmonary capillary pressure, the net force would still be for reabsorption rather than filtration and edema formation.

58 A patient with hypertension accompanied by increased total peripheral resistance and normal cardiac output and stroke volume will have an elevated diastolic pressure but not an elevated systolic pressure. True or false?

False. With a normal stroke volume, an increase in total peripheral resistance will cause a proportional rise in both systolic and diastolic pressure (Chap. 7).

59 Why would renal artery stenosis produce hypertension?

Because of the high resistance of the stenosis and the pressure drop across it, glomerular capillary pressure and therefore glomerular filtration rate are lower than normal when arterial pressure is normal. Thus a renal artery stenosis reduces the urine output rate caused by a given level of arterial pressure. The renal function curve is shifted to the right and hypertension follows.

Suggested Readings

GENERAL

American Physiological Society: *Handbook of Physiology*, sec. 2: *Circulation*, vol. 1: *Heart*, R. M. Berne (ed.), 1979; vol. 2: *Vascular Smooth Muscle*, D. E. Bohr (ed.), 1980; vol. 3: *Microcirculation*, E. M. Renkin and C. Michel (eds.), 1980. Distributed by Williams & Wilkins, Baltimore.

Berne, R. M., and M. N. Levy: *Cardiovascular Physiology*, 3d ed., Mosby, St. Louis, 1977.

Guyton, A. C., A. E. Taylor, and H. J. Granger: *Circulatory Physiology II: Dynamics and Control of the Body Fluids*, Saunders, Philadelphia, 1975.

Guyton, A. C., et al. (eds.): *International Reviews of Physiology. Cardiovascular Physiology I, II, and III*, University Park, Baltimore, vol. 1, 1974; vol. 9, 1976; and vol. 18, 1979.

Little, R. C.: *Physiology of the Heart and Circulation*, Yearbook Medical, Chicago, 1977.

Rushmer, R. F.: *Cardiovascular Dynamics*, 4th ed., Saunders, Philadelphia, 1976.

Shepherd, J. T., and P. M. Vanhoutte: *Human Cardiovascular System: Facts and Concepts*, Raven, New York, 1979.

Smith, J. J., and J. P. Kampine: *Circulatory Physiology–The Essentials*, Williams & Wilkins, Baltimore, 1980.

To keep abreast of continuing developments in cardiovascular physiology, interested students should peruse the following journals: *American Journal of Physiology, Circulation Research, Journal of Molecular and Cellular Cardiology, Acta Physiologica Scandinavica,* and *Pfluegers Archiv.* In addition, *Physiological Reviews, Circulation Research,* and the *New England Journal of Medicine* all publish excellent review articles on cardiovascular physiology.

CHAPTER 1

Johnson, P. C.: "Renaissance in the Microcirculation (Brief Review)," *Circ. Res.*, vol. 31, 1972, pp. 817–823.

Landis, E. M., and J. R. Pappenheimer: "Exchange of Substances through the Capillary Walls," in *Handbook of Physiology*, sec. 2: *Circulation*, American Physiological Society, Washington, D.C., 1963, vol. 2, pp. 961–1034.

Renkin, E. M.: "Multiple Pathways of Capillary Permeability (Brief Review)," *Circ. Res.*, vol. 41, 1977, pp. 735–743.

Starling, E. H.: "On the Absorption of Fluids from the Connective Tissue Spaces," *J. Physiol. (London)*, vol. 19, 1896, p. 312.

Zweifach, B. W., and A. Silberberg: "The Interstitial-Lymphatic Flow System," *Int. Rev. Physiol. Cardiovasc. Physiol. III*, vol. 18, 1979, pp. 215–260.

CHAPTER 2

Fozzard, H. A.: "Heart: Excitation-Contraction Coupling," *Annu. Rev. Physiol.*, vol. 39, 1977, pp. 201–220.

Lindsay, A. E., and A. Budkin: *The Cardiac Arrhythmias*, Yearbook Medical, Chicago, 1975.

Trautwein, W.: "Membrane Currents in Cardiac Muscle Fibers," *Physiol. Rev.*, vol. 53, 1973, pp. 793–835.

Weidman, S.: "Heart: Electrophysiology," *Annu. Rev. Physiol.*, vol. 36, 1974, pp. 155–169.

Wiggers, C. J.: *Circulatory Dynamics*, Grune & Stratton, New York, 1952.

CHAPTER 3

Bishop, V. S., D. F. Peterson, and L. D. Horwitz: "Factors Influencing Cardiac Performance," *Int. Rev. Physiol. Cardiovasc. Physiol. II*, vol. 9, 1976, pp. 240–273.

Brutsaert, D. L., and W. J. Paulus: "Contraction and Relaxation of the Heart as a Muscle and a Pump," *Int. Rev. Physiol. Cardiovasc. Physiol. III*, vol. 18, 1979, pp. 1–31.

Index

Page numbers in **boldface** type indicate figures.